The Road To Find Out

William J Wilkins

Published by William J Wilkins, 2023.

Jim & Mary Ellen,
Thanks for
Reading!
Bill

THE ROAD TO FIND OUT

First edition. May 29, 2023.

ISBN: 979-8223135531

Written by William J Wilkins.

... to the kids who know they're not there yet.

Overtime

This November morning was barely an hour old.

The Emergency Room's steel furnishings looked as stark and sterile as the fluorescent light that flooded it from overhead. An hour earlier I had awakened with an odd tension in my chest. Neither antacid tablets nor food would make it stop, so I headed for the hospital wondering, with my family history, if it just might be my heart. "It's probably nothing," I said to myself, "but what if it isn't nothing?" So, before I got to the corner, I turned back to get my wife. Together we continued the 10-minute drive to the hospital – me still behind the wheel. That move, we would soon realize, was absolutely a mistake.

In the ER, they gave me nitroglycerine. That made the feeling back off instantly. A heart attack was imminent, they warned. The doctor glanced at his watch and then swung his gaze to me. "I'm not going to tell you what to do," he cautioned, "but if you were my dad or my brother, I'd direct you to take a ride downtown to the heart center. This time, though," he said with an eyebrow raised, "we have a driver for you." He looked down at the floor, then back up at me. Suddenly deadly serious, he added, "It's your call, but if I were you, I would not wait."

His words just hung in the air.

I calculated that I had just blown a perfectly good night's sleep and was already running up a very unhealthy hospital charge. My wife Megan nodded encouragement for me to take the doctor's advice. Going back to bed was apparently not going to be one of my wiser options.

I could always sleep a different night?

Never having been in the back of an ambulance, I'd have to say it was not what I anticipated. There are no windows and you can't even really see the paramedic who's riding with you. You look up at the steel ceiling. You feel vulnerable. You think about what could be coming next. I told the emergency tech that I imagined he must see lots of frightened people on his overnight shift. He affirmed, but did not say more. I glanced over. He was intently monitoring my vitals.

When they wheel you into a hospital on a gurney, you hear the voices of the care providers but you don't see them. Instead, you see different ceilings moving by, stopping, sometimes turning, then moving again. You eventually arrive somewhere but you don't exactly know where that is. For me, it turned out to be room #6842 in a large cardiac care hospital in Grand Rapids, Michigan.

When the coronary specialist walked in at 3:30 a.m., he was not so sure I should be going home. He began very deliberately, "Now, I just want you to know that now that you are here, nothing is going to happen to you." That sounded like something they say to people who are in for some real trouble. I must have had a perplexed look plastered across my face.

He continued, "One of the three primary arteries in your heart is 99% blocked. We have to get it opened now or it's not going to be good." Through my grogginess I realized he was saying that unless I followed the program he was about to outline, I might not be getting more sleep – ever.

Now I was fully awake.

"You will go into the 'Cath Lab' as soon as it is available ... and the procedure takes about an hour. We are going to put a stent in the blocked artery to open it up. You can't eat until after the procedure but we can give you some ice chips." Then he nodded, "Any questions?"

All I had were questions. I had way too much time to think. So, despite my best efforts at exercise and healthy eating, was I following in my father's footsteps, his father's footsteps and his father's fathers? I had always hoped that because I had never smoked, I could duck my family's curse.

The disappointment was overwhelming.

At 1 p.m. a new shift nurse arrived with news that a big emergency had come in and there would be a delay with my procedure. Then at 1:30 the nurse's tone changed. "We need to get you into the lab soon. We are going to try to move you up in priority."

Megan began to look more worried. I could feel the tension increasing again. Was each heartbeat a gift or a given? How many did I have left? I had always taken for granted that my heart just did its beating thing automatically. But there must be a finite number of beats for every heart, and the final one for mine apparently could come at any moment.

About 4 p.m. the ceiling began to move again. This time I saw a doorframe go by and I was able to glimpse the hallway that lay beyond. There were doors as far as I could see. "What floor am I on?" I wondered aloud.

"Six," said the matter-of-fact young orderly pushing my bed on wheels. "You are in room number 6842." I considered how massive the hotels are that have rooms numbered in four digits. While cloistered in my cube I had not given a thought to the immensity of the facility that surrounded my little bed. I was but one life in a building of thousands.

With little else to think about besides ice chips, the number "sixty-eight forty-two" began to take on an undue significance. I turned the numbers in my mind. It occurred to me that 68/42 was actually a familiar pattern. "US 68" is the highway that makes its way north from the southern Ohio foothills to the flatlands, exiting the state at Toledo. It's a north-south route for the locals. "US 42," on the other hand, is a diagonal thoroughfare, traversing Ohio's largest cities: Cincinnati to Columbus to Cleveland. The two highways intersect right in the middle of Xenia, Ohio, the little town where I was born. "X marks the spot," they always said of Xenia, but I had never thought of it in quite *that* way before.

The administrator stopped in and was careful when she advised, "You do understand that you do not have a 100% chance of a successful outcome today?" Translation: "This procedure we are about to do could kill you." I acknowledged, and signed away my life on her clipboard.

Then a pleasant anesthesiologist walked up and introduced herself and her team. She graciously apologized for the delays. After explaining what was about to happen, she took a moment to ask where I was from. I told her, "I'm not from here in Michigan. I was born in Ohio."

"Oh, I was born in Ohio too," she commented, "but it was in a little town I doubt if you've ever heard of." I looked up at her just as she was administering the anesthetic. "I was born in a little town named 'Xenia,'" she said with a smile as I was drifting away.

Back in my room with the anesthesia wearing off, I wondered if she had really said what I thought she said. But as the fog lifted, I became certain that my encounter with the woman who hailed from my hometown was quite real.

A coincidence? As I reflect on it, I think not.

I left that hospital with the distinct impression that someone, somewhere had known all along that I would nearly expire on that November day. And that someone also knew that I had begun my days ... but would not end them ... at 68 and 42.

If you get uncomfortable when you hear people talk about ghosts, UFO's, premonitions or poltergeists, you may not want to read any further. None of those are in this book, by the way. But the point is, if you have the kind of mind that needs to fit all happenings into neatly defined spaces, you may want to stop. You may not like what's coming. Some of the things that you are going to read about – many of which have happened to me personally – are not easy to categorize. Instead, if you are an individual who is willing to consider that there just might be aspects to life that cannot be readily seen, heard, touched or understood, I encourage you to continue.

Is it novel to compare one's life to a journey or a road trip? Hardly. But I submit to you that our existence here consists of not one road, but two. The one that we all know best spans the physical landscape. It's the one measured by equally-spaced mileposts. But intertwining is another path that weaves in and out. Distance on this avenue is measured in moments, not miles. The Road to Find Out is not a physical road, yet it is real, nonetheless. Invariably, it leads some souls to wisdom and others into a wasteland. Why?

Like you, I started my trip years ago. But before long, I found myself sorely in need of a guide. At the time, I thought my wanderings were an invaluable opportunity to explore. In hindsight, though, a focus on what is true would have been a smart trade for the unfettered freedom I chased.

Truth is what supersedes the subjective. It's what I began to long for when, as a boy, the adults would not – or could not – answer for me the basic questions of living. It's probably obvious that I was left to find many of the answers on my own. In the end, that turned out to be okay. But there have been way too many detours and wasted hours. Maybe you know what I mean.

To paraphrase songwriter Bob Dylan, "if you've got nothin', you've got nothin' to lose." By extension: a person in his last days is someone who can be trusted because he has no reason to traffic in untruth. I could easily have expired on that November morning. A couple decades earlier, I definitely would have. So, understand that I have zero interest in fabricating fiction as I complete my time here. I am aware that this writing may not gain me fans in certain religious, academic, philosophical or secular circles. And you may have to put this down at some point if you cannot accept what you find. But I write to you the unbridled truth about what I have experienced and concluded.

Much of it – like my hospital experience – is impossible to categorize.

When I told my longtime hiking friend about this project, he listened intently then told me of an occurrence that his own family had experienced. He grew up in the house right down the street and in his front yard was a lamp post with a sign with his dad's name "C.J. Wendell." I recall that black sign with the white lettering hanging there the entire time I lived on that street. It swung from two hooks on a horizontal bar. After residing there 43 years, Mr. Wendell died. A funeral service ensued, and upon returning to the house afterward, the family was shocked to see the old sign hanging askew from only one hook. On close examination they discovered that one of the hooks had just rusted through, purely from age. It had held the little sign with Mr Wendell's name for 43 years then broke in half *during* his funeral service.

"How can you explain that?" my friend asked me. His question begged for an answer.

In another instance, a coworker of mine recalled that he had gone on a crazed road trip after resigning his broadcasting job. After many days in a liquor fog, he awoke in the back seat of his car. The floor was littered with booze bottles and pizza boxes. He sat up and realized that he had ended up in the parking lot of some large professional building. Disheveled and dizzy, he walked in the front door to ask to use a telephone. The lady behind the counter saw him coming, and assuming him to be a new patient, checked him into the Tampa Florida Alcohol Rehab Center. He ended up staying there for months and got clean. He swears to this day that this inexplicable experience was no coincidence.

"What it was that brought me through their doors, I'll never know," he asserts. "But it certainly was no accident."

There's a reason that you specifically have this writing in your hands right now. I have become convinced that nothing ever happens by accident. And the agent responsible for occurrences like these? That is what this is about. It's a mystery that confronts you and me and everyone alive. And though it will not be resolved as I complete my overtime period, I will continue to try to get my hands around it.

We will.

The Reality We Are Presented

Sometimes Truman's life almost seemed like it was scripted.

That is because it was.

Perhaps you have heard of *The Truman Show*[87] - a movie in which Jim Carrey played a man raised on a TV sound stage constructed to look exactly like a small New England town. What Carrey's character was never supposed to know was that he was actually the unwitting star of a television program. The stage was the only reality he was ever intended to experience. The show was his life – literally. His family and friends were all actors. While viewers watched through disguised cameras, the show's producers manipulated an unsuspecting Truman, the main character in a nightly "reality" drama, watched by millions.

The creator/producer of the program was a man named Christof, who, in one scene, explained to viewers that forcing Truman to live his entire life as a specimen was not immoral. Christof defended the program, affirming the film's premise, "Actually, we all accept the reality we are presented."

Meanwhile, Christof's specimen was beginning to ask increasingly awkward questions. One day he recognized a homeless man on the street who looked exactly like some old photos of his stage "father," a man who was supposed to have died in a boating accident when Truman was a baby. As he observed the man, Truman could tell it was the very same person. How was this possible?

The glaring incongruity led Truman to begin thinking outside the box – literally. He became suspicious of his "reality" and began quietly testing boundaries for inconsistencies. Were his family – and the people he had known all his life – merely play-acting? For what conceivable purpose?

Convinced nonetheless that he was being manipulated, he secretly began putting clues together. Eventually he found a way to escape the director's sight and confirmed that the world he had come to accept as reality was actually a clever fiction; the unseen world just outside the set was indeed the one that was real.

At one notable point during this progressive revelation, Truman became attracted to one of the show's actresses. She was just an extra, but Truman's infatuation was real and that became a real problem for the writers because their budding relationship was not in the script. His bride-to-be was intended to be a different woman – already preselected. Quickly, the show's directors, dressed as townspeople, descended on the pair to distance them. It was at the last possible moment that the actress blurted, "Don't you realize that this is just a TV show, Truman!?" She screamed at him, "I am just an actress!"

The woman was whisked away by a sea of pseudo-pedestrians and was never again seen on the program. Curiously, all evidence of her was soon erased – even her name and her phone number disappeared from the "town" phone directory. But Truman would not forget. While he could, he located what physical evidence he could find to prove her existence and even squirreled away a photograph that he had taken.

Truman had briefly tasted the truth. Now, his choice was a stark one: he could continue the pretense, knowingly living in a comfortable fiction, or he could make the leap into a new world of uncomfortable, but very real, unknowns. In the end, he chose the leap, escaping the confines of his handlers.

No one was happy when the show ended – not the producers, the actors or the viewers – only Truman. He located the actress and they began a real relationship in the real world.

Through it all, Christof's premise proved wanting. Not all of us accept the "reality" that is presented to us. We test it. We try it. Ultimately, we accept it or we expose it.

Perhaps like Truman, you have begun to suspect that there's more going on beyond the horizon, just outside your field of view. In my experience, odd occurrences like the one in the hospital have shown up at times of special emphasis, positioned along the road like signposts. For me, they have not come calling at random. Instead, they have been more like the extra, shouting at critical moments, "Don't you understand that reality goes way beyond what you can see?"

No, you and I don't have secret cameras trained on us. It's actually worse than that. If that thought unnerves you or intrigues you, it could be you are approaching a point on your journey where your full attention is required.

Willie, Billy and Bill

This story really began in May of 1942. Willie woke to a sharp pain between his eyes. He felt as if a vice had been squeezing his face. He rolled over to hear his wife say, "It's bad again, isn't it?" Louise had been watching him struggle to sleep. He didn't need to answer. Shuffling to the medicine cabinet he found an almost-empty bottle of aspirin – the only thing that would blunt the throbbing from his chronically infected sinuses. Willie and Louise decided that morning that he must try a new operation that their family doctor was recommending.

It was a decision that would only bring more pain.

Just before Christmas their doctor had told Willie about a new procedure that allowed sinuses to be drained through vents drilled upward from inside the mouth, just beneath the cheek bones. It sounded awful but it offered hope of relieving the malady that had increasingly made Willie's 61 years miserable.

The late spring sun was setting over the river flatlands as they left Arkansas and crossed the spider-like Mississippi River bridge to the west side of Memphis. Daughter Jessie and her husband had left work early to drive them to Methodist Hospital for the procedure, which doctors said could keep Willie there a week. As the car crested the span, Willie looked back at the sun hanging low over the cotton fields. It had been almost 40 years since he had followed his friend Arthur across that very bridge from Tennessee to the cotton oil plant in Forrest City where he had come to occupy the superintendent's chair.

Willie was the first in his family to make a living from manufacturing instead of farming. His life had been a success by all standards. Willie and Louise fairly knew everyone in their town and the oil plant job had grown into a lifelong career. Yet, the time had not been without difficulties. Perhaps the most pressing was the new war that was embroiling the country. The Pearl Harbor attack had stunned the nation just before Christmas and the whole family was concerned because only-son Billy was old enough to fight. In fact, he could not accompany the family to the hospital that evening because he was away at Army Signal Corps training in New Jersey, anticipating his deployment.

Billy was an enigma. The awkward redhead was the family geek, pursuing his fascination with electricity and radio waves. He spent his days with his nose in technical books and read so many titles that the town librarian would call the house whenever a new one arrived.

Willie and Billy had never been close and Willie suspected it was due to his long hours at the plant. Billy had rejected the Methodist Church and everything that Willie called near and dear. His only son seemingly had cast everything aside when he departed for college. He even left his name. No one would know him as "Billy" at Mississippi State. He made sure of that. Now, as Willie gazed across the valley, more than anything he wished he could return to a day when he could still engage Billy, father to son.

Perhaps it would have comforted Willie to see that he and Billy actually shared a lot. Willie had left the farm life to become a factory manager in a new industry called "manufacturing." Billy would leave the small town for the big university and would pioneer a career in an industry that would be called "electronics."

Arriving at the hospital, Willie and Louise settled in for the night in preparation for the morning procedure. A general anesthetic would be required, and in 1942, that was a very new concept. No one they knew had any experience with being "put to sleep" as the doctors called it. Louise opened Willie's *Bible* and handed it to him to read as he got comfortable. Louise found the Memphis newspaper and surveyed spring fashions. The room telephone rang and Jessie was on the line, already back at home, checking in.

Willie's mind drifted to his own father and mother, Joe and Belle, buried back in little Springville, Tennessee. What would they think about all this technology? He recalled vividly his mom screaming from the next room, giving birth to his baby sister. He also remembered the silence. Had there been "modern medicine" in rural Tennessee in 1890, it might not have gone the way it did for Belle and the baby who never got a name.

Louise used her sewing scissors to snip a poem out of the newspaper that she thought Willie might appreciate, and he did. He had just finished reading it as he nodded off. She used the clipping to mark the page in the *Bible* that he had been reading, then she quietly closed the book.

...to be continued.

I met Billy in 1955. More accurately, he met me. He was my dad. He had returned from the war in the late '40's, having done his Signal Corps hitch on the island of New Guinea. I never heard anyone call him "Billy" unless his mom and sister were visiting.

One thing he did seemed to have brought back was a penchant for travel. I noticed that when we were on a vacation trip, unless we had arrived at our final destination, we never stayed more than one night in the same place. When the sun came up, we packed up.

But I'm really not sure if it was the war that generated his drive to be in motion. Perhaps the tendency seeped down through the generations. Turns out that neither Billy nor any of his ancestors going back ten generations are buried in the same county where they were born – and only a few are buried in the same state. Dad's family came to America in the 1600's and lived in the Colonies. They were all about horses: breeders, horse borne dragoons in the Revolution and circuit-riding pastors in the Tennessee hill country.

My mother Rose's family came in from Hungary early in the 20th century, settling in the Northern Minnesota Iron Range. She would leave her little hometown for the likes of New Jersey after college where my sister would be born. So, the inclination to roam came at me from all directions. In fact, in our little family of four, all of us were born in different states.

I grew up thinking that was normal.

I was exposed to our family wanderings at an early age. Initially my parents and sister would drop me with my grandmother in Arkansas while they took a summer vacation. But in 1961, without a lot of explanation, I found myself a passenger on what seemed to be a never-ending journey to the West Coast. The reason it seemed so long? I was six years old. Childhood makes everything seem oversized. Remember your giant neighborhood or the massive house or apartment you grew up in? This 3-week trip became an epic saga in my first-grade psyche. Somewhere along the way I think I resigned myself to the possibility that I might never again see home. And, of course, at six you can barely see out of the car window. The turquoise vinyl interior of a Chevy Impala, no matter how in-vogue, can only hold a six-year old's fascination for so long.

But there were two other reasons for the trip's seeming endlessness: highways and heat. In 1961 the Interstate system had only been under construction for a few years and still had a very, very long way to go. We would take the ramp onto a new interstate highway section and roar at 70 mph for 10 minutes before, "Divided Highway Ends." It was back to two-lane roads and local traffic ... again and again ... and again. In some places in the Midwest, Interstate 70 was being built right over the top of the original National Highway - US 40. Farther west, what would become Interstate 40, scarfed up much of Route 66.

In short, it wasn't short. But no one knew that at the time. My parents marveled at the high speeds they could briefly achieve on the "modern highways" and commented repeatedly how long the trip would have taken in the "old days."

On the very first day, heading west from Ohio, Dad pointed out two massive construction cranes that were assembling a pair of what appeared to be buildings just west of the Mississippi River at St. Louis. "Someday those two 'buildings' will meet hundreds of feet in the air in what will look like a giant arch," he said.

But I could not imagine an arched edifice at six. All I knew was that as we proceeded west it was getting hotter every day. I would have longed for an air-conditioned car except that, well, I had no idea that such a thing existed. Plus, that was not an option on the Chevrolet Impala in 1961 – at least not *our* Chevrolet Impala. No family I knew had an air-conditioned car. And no one I knew was taking a driving vacation to the West Coast either, except that we were because *we* did that kind of thing. Our intrepid family traveled at all costs.

The 1961 challenge: pull Dad's homemade "pop-up" trailer to something called Disneyland which had just opened in a place called Anaheim. That's in California. The prospect of a 3000-mile trip towing a hillbilly camper certainly didn't daunt the Wilkins family. I vividly recall the pungent aroma of that hot tarp flapping in the desert wind as we foisted it above the plywood trailer frame each evening, the varnish still sticky.

Yes, it would be another night to remember.

Neither did the prospect of crossing the Mojave Desert give us pause – apparently – because we took it on in July, no less. And we beat it back gallantly with no more than a plastic tub of ice water and a wash rag.

Yes, it could get worse, actually.

Our Walley World saga totally went off the rails in the Golden State when Mom drove the car into a campground sign post and Dad needed a midnight hospital run to pass a kidney stone. Amazingly, both the repair to the car and to Dad added only one week to our trip.

For those who could actually see out the window, it must have been a happy moment when they saw the sign that read "Welcome to Ohio."

In hindsight, my sister's contribution to the trip was most welcome. In the early 1960's, it was still possible to travel coast-to-coast and log hundreds of mom-and-pop motels, restaurants and gas stations as you snaked in and out of tiny towns on the state roads. Through it all, her spiral notebook dutifully documented every ice cream joint and dairy bar – no matter the size. True, there were already about 3000 Dairy Queens around by then. But it was the "mom & pop" competitor – with a unique flair – that was the sought-after find. The Dairy Freezes, Polar Cones and Penguin Points were most prized. Even the smallest towns had an entry because back in that day, each motorist was required to parade past a gauntlet of traffic lights, neon signs and fast-food stands to make it back out to the highway.

When we arrived home, that impressive list of ice cream stores filled many a sweat-wrinkled page. The excitement of those tasty finds animated what would have been many longer hours. Of course, we didn't stop at each store. We had a schedule to keep.

But my sister's greatest contribution to my life on that trip was when she bet me $50 that I would not make it to the age of 21 without becoming a smoker. [$50 then equates to more than $250 in twenty-first century money.] It was a shrewd wager on her part since I still had 15 years to go before my 21st. She was already smoking by 18 and it probably seemed to her unlikely that I wouldn't take up the habit by the time my teen years were over. And just to pump up her odds, she defined "not being a smoker" as not smoking even *one cigarette.* Not one.

I remember thinking that to win that bet was going to be very, very easy. It only required me to do nothing. I made sure to collect all $50 from her the day I turned 21.

That west coast marathon marked a turning point for this writer. Sister soon went off to college and no longer was I off-loaded in

Arkansas while the family embarked on their summer tour. And the rides got more interesting as I got older... and taller. In 1964 we traveled to the New York World's Fair. Three years later it was up to "Expo 67," the exposition in Montreal. In between, there were family visits to Arkansas and Minnesota. For me, between the Razorbacks and the Lumberjacks, it was a full-on culture whiplash. And somewhere along the way, I was reassigned to be the front seat map-reader. Mom was only too pleased to read books in the back. But it would not be long before my mother would be required in the front seat again – this time to take the wheel.

The Man Next Door

The name "Xenia" comes from a Greek word that means "place of hospitality." The little town with that name appears to have been just that for my family, who settled there after my dad returned from the Pacific to his assignment at Wright Patterson Air Force Base. It's in the center of what they call Ohio's "Transportation Triangle" formed by Interstates 70, 75 and 71. He was the second of three Williams and I was the third, and it turned out that we three had more in common than the name. Each of us had but one son – the youngest child – and each son was born the year the father turned 37. So, when I turned up at Greene Memorial Hospital in Xenia in 1955, it should have been no surprise. My father turned 37 that year.

We lived there from the 1940's to the 1960's. Then, in the fall of 1965, we moved ten miles west to Kettering, a suburban area that had been stretching out across the hills south of Dayton ever since Charles Kettering developed his automobile electric-starter there.

In fact, there were so many "firsts" in the Ohio that I came to know, no wonder it seemed nothing was impossible to create – or by extension – to repair. The Wrights had designed their first plane at their bicycle shop in Dayton and practiced flying it on the farmland that would become the Air Force base. Inventor Thomas Edison (light bulb, phonograph, movie camera, generator) was from Northern Ohio. The Crosley's pioneered radio broadcasting less than an hour south, near Cincinnati. John Glenn, the first American to orbit the earth was born in Cambridge, OH. Astronaut Neil Armstrong grew up just an hour north in Wapakoneta. It was he who would first walk on the moon in the summer of 1969.

It was that same summer I discovered a harsh reality: despite the staggering progress of the 20th century, there are certain things that are beyond the control of even our most learned. I was two weeks out of 8th grade and understandably geeked about being on summer vacation. It was June. The moon landing would happen in less than a month. Our family was planning a trip to Florida that August. But as it turned out, Neil Armstrong's trip came off exactly as planned – but ours never did.

On the evening of the 19th my mother was at the store and I was in the back yard playing ball with the neighbor guys. I heard my father's voice calling me from across the yard. There was something about his tone. He was seated on the couch looking at me through the sliding glass door as I approached. I pulled it open and asked what was going on.

He inquired if I would mind staying in the house with him, saying the he needed me to get him something to soothe a stomach upset. I thought it odd that he didn't get it for himself. After he took a few tablets, he laid down on the sofa. I had never seen him do that. At a bit of a loss, I sat down in his chair to watch a rerun that was just finishing on TV.

As 9 p.m. approached, the sun was touching the horizon and the final credits were rolling. I looked up and Dad was still laying quietly. It was getting dusky and someone needed to turn some lights on. He was not getting up to do it. I glanced out the glass door and my friends had all gone in. I looked back. Something was wrong.

My recollection of those moments is indelible, yet blessedly inaccessible.

Dad motioned to me, and in a quiet voice said, "Call the emergency squad." I wondered if he was signaling me to fetch something for him to get sick in, given his stomach. The little lavatory adjacent to the family room had a trash bucket. But when I returned an instant later, he no longer knew I was there.

What I watched unfold wasn't anything like a stomach upset. I ran to the kitchen phone and dialed "O". [In those days the local operator would connect a caller to an ambulance service.]

"What is your emergency?" came the voice.

"It's a heart attack," I blurted. It was a complete and total guess, but it was the only thing I could think of that would make an ambulance come fast. I gave the lady the address and hung up. I turned back to my dad but it was useless. I recalled film strips from health class where you save a drowning victim with mouth-to-mouth methods. "That's not going to work here," I remember thinking. What I was watching was nothing like the film strips.

So I ran to get my friend's dad who lived next door. I didn't even knock. Throwing open the front door of his house, I called out something that made him come running behind me. The man took one look at my dad and put his arm firmly across my shoulders. He walked me into our living room where I couldn't see.

It was the right move. And I did stay there for a few moments. But in the end, I came back and stood with the man, now both of us watching. It was like a bad accident scene. We couldn't do a thing. Then I just couldn't watch any more.

I walked out the front door to wait for the ambulance but what greeted me on the porch was silence. I stood awkwardly, in full view of the neighborhood, the door ajar behind me. Finally, there was a faint siren, which became louder until I could hear its wail and see its red-and-white lights sweeping across the homes on the main road. The ambulance had stopped in the intersection and was about to come down our street. But it wasn't coming. I stepped into the yard as if to put my arms in the air, but the driver was not looking. A neighbor who was watering his lawn looked up.

The ambulance began to turn but my anticipation turned to horror as it turned the wrong direction. Instead of headlights, now there were tail lights.

Now the neighbor was walking toward me.

Then, at the very same intersection, my mom's car appeared, returning from the store. She glanced up with a curious expression as she approached, spotting me in the yard. I think it must have been the man next door who met her in our driveway before she could encounter the scene.

Big pieces of that night have gone missing. All I recall clearly is that the paramedics found the house and they got an apparatus strapped over my dad's face to help him breathe. With one man in front and one behind the gurney – followed by the family – we filed through our dining room toward the front door. The family was forced to hold up as the paramedics turned the gurney 90 degrees in the small foyer. The maneuver took a few moments with equipment clicking. Oddly, this allowed my father's form to pass full length before us all as he was wheeled out. I looked down to see his face, contorted under the plastic mask that was pushing air in and out. One of the paramedics consoled my mother that he was still breathing; that he was still alive.

But something advised me otherwise.

Mom rode in the ambulance. I was with the others, squeezed hastily into someone's car. On the way to the hospital, I wondered if I should say something. The ER doctor took the family into a consultation room and told us that there had been no way to save him. He was saying that it was a massive heart attack and there was nothing that could have been done.

He said that to make us feel better, but it didn't.

It was dark when we returned from the hospital to begin calling family. Then, sleep would not come. One by one, each of us showed up in the very room where he had breathed his last. We talked all night. No one ever turned on a light.

That all happened decades ago, yet the event still reverberates in me today. The shock of his abrupt departure was amplified because he was relatively young and considered healthy. There had been no doctor warnings; no chronic disease. "He was just 50." That would be the mantra at his funeral. But the more they railed against it, the reality just became more indelible. This break would not ever be repaired. He was now permanently absent.

My 14-year-old mind could not wrap itself around it.

Sometimes during my teen years, I would sit in that room alone and ponder. Would he do something to show me that he was watching? Could he silently move a vase to prove he was still there? Of course, nothing like that ever happened. But the message on that June night was unmistakable and I continued to be shaken by the knowledge that someone, somewhere knew that my dad had passed from the bonds of life before any of us knew – and before any of the doctors. How? I carried these questions with me throughout what remained of my adolescence, but I dared not express them. "Clearly, Ma'am, your son has experienced a trauma and perhaps even has had a break with reality. It's not unheard of...."

I began to tell my mother about it all once, but the effort came up short. The phone rang and I never again broached the subject. No, I didn't hear my father speak to me. No, I didn't hear a voice. And yet, someone uninvited knew that he was gone.

After the funeral was over, family left and the first few months passed. Personally, it was more often in the unguarded moments that I would feel, not Dad's presence, but his absence: a half-finished project on the workbench; an item of clothing that still hung in the closet. Through it all, Mom and I tried to behave as we always had. I never shared with anyone all the things that happened in those awful minutes. Yet, now with the passing of time, I can better see how the night of the nineteenth seared me. I grew up intensely bearing the questions that I absolutely knew must have answers, but which absolutely could not be asked.

The time passed as it always does – one day at a time. High school days were drawing to a close. My brown sport coat, tan shirt and diagonally-striped super-wide tie hung neatly on my closet door, looking at me through a plastic drycleaning bag. When someone walked in or out of my room, the bag would ripple silently, ever so slightly. It was nearly senior picture time. It seemed that once that outfit came out of the bag I would be finished with my youth and ready to go somewhere to start being an adult.

When the day came to get dressed, I stepped slowly down the hall stairs and into the kitchen. My mother turned to admire my spit-polished, high-school senior self. "Now don't you look nice," she complimented. My tie was still loose about my neck. Mom glanced toward the phone and placed a call to the man next door who agreed to come quickly and tie it for me. No one needed to say more. Dad would have done it had he been there. The photo came out okay and I stepped into "adulthood" without further incident.

Then many more years passed. I occasioned to visit Aurora, Illinois one winter day for a business meeting. The timing was early, so I stopped into the hotel's breakfast room for a quick coffee and cereal.

That's when I saw them.

I was attending to my breakfast, but each time I glanced up, the dark-haired woman two tables away would move her eyes ever so slightly. It seemed she was watching. She was huddled with a young boy as she fussed quietly with his clothing.

As soon as the man at the table between us departed, the woman stood up quietly. I thought she and the child were about to leave, but instead she wheeled to face my direction. They were both dressed in black. The boy was perhaps six. Tugging him by one hand, the lady walked in a deliberate manner directly to my table.

She stopped and just stood for a moment, looking down at the floor. It was awkward in the extreme. She did not introduce herself or say hello. The boy looked up at her, but she did not look back. Then, she seemed to gather herself, leaned toward me, and in a terse whisper uttered, "We are here for the funeral of the boy's father. His tie needs to be tied."

It was not a question. It was not a request. It was a statement. His tie needed to be tied.

I was in a hurry that morning. Tying the boy's tie was not my job. There were plenty of reasons I could have declined to help. I just couldn't think of any good ones. Meanwhile, the woman and the boy waited.

Time ground to a full stop at the Comfort Inn.

Tying my own tie is hard enough, sometimes. Tying someone else's is just backwards and, well, it's really awkward. It's not like shoe laces because there is a left-side and a right-side to a tie operation which makes tying someone else's tricky.

I fumbled. I am sure I did not get it on the first pass or the second. But with the other diners glancing up from their waffles and probably trying not to stare...I was eventually able to make the tie look passable.

I looked up at the woman. She only nodded, then buttoned the boy's coat. Without a word, she turned and took his hand and they walked out the hotel's front door into the cold Illinois morning.

For the hotel guests with whom I shared the breakfast room that day, perhaps it was just an unusual moment. But for me it was a not-so-gentle nudge. I left the hotel that morning knowing that it was time for me to stop acting like the boy with the widowed mom.

It was time for me to start acting like the man next door.

House Afire

Nature, they say, abhors a vacuum, and into the wide-open space left by my dad's departure, entered the closest explainer our family had – the Catholic Church. On a sweltering August afternoon in 1969, one kindly Father Garrigan arrived at our home in Kettering. He was not just a priest. Father Garrigan was a "monsignor" – a priest's priest. He had a special rank in the priesthood, and from what I could discern, was both respected and liked at St. Christopher's Parish – the Catholic franchise where we attended.

When the big day dawned, it was apparent that our family was hosting a dignitary. Mom tidied up all morning, scurrying room to room. The good father didn't exactly arrive in a motorcade, but it was a rather large, freshly-washed black sedan that rolled serenely up our suburban driveway. The sun reflected off its hood and projected an irregular pattern of light into our shady back porch. Inside was Garrigan with another man who doubled as his driver. In their crisp black attire, the men stepped briskly across the steaming black asphalt. The hot wind whipped at their pant legs as I watched them approach. I could see the men talking, but through the sliding glass door I heard only silence.

I suspected this would be the closest thing to a "Papal Visit" our family would ever experience.

Inside, our guests were offered refreshments and then ushered into the living room where they took their positions; Garrigan and my mom were seated front-and-center on our family's "virtually never sat-on" white couch. Flanking, in fuzzy fabric chairs were my sister and I, and of course, the driver. It was the middle of the day, and I remember thinking that in this brightly lit room with its white-painted walls, the scene took on a "cloud-like" appearance. Mom and the Monsignor, both in black, faced each other as much as you can on a couch – or on a cloud.

I recall Garrigan reassuring my mom in measured tones, that she could be certain that my father awaited her in heaven by virtue of the fact that she had all the appropriate credentials: she went to mass weekly; she took the sacraments; her confessions were up to date.

It turned out that this meeting indeed would be a black and white moment for me. My father, who had shunned and even mocked religious practice, was being involuntarily ensconced in high places because of Mom's credentials at our church ... in Kettering, Ohio, no less?

It was really too much.

Later, I went to my room and thought about it more. I liked Father Garrigan, but his words rang hollow in my ears. I believe that the moment that he told my mom that she had punched my dad's ticket to paradise without his consent, I set my face to find out what is really going on in these matters of life and death ... and more specifically, if there is anything to life *after* death.

The Monsignor's words may have calmed my mother's fears but they only stoked mine. Deep down, I knew I had to get to the objective truth about this existence thing because privately, I suspected that someday I might ply the same path as my dad physically, and perhaps spiritually, too. I vowed in my heart of hearts not to take that final step into the next world unprepared, as I suspect he did. Even if I had to follow him in sudden heart failure, I would not follow him in spiritual ignorance. My dad did not believe there was a life after death, so it was pretty evident to me that if there really was such a thing, he had no idea what lay ahead.

The prospect was terrifying.

His demise was startlingly abrupt. When one realizes that the end may be moments away, what is the content of one's last fragments of thought? Does a person yearn to say goodbye? Does one grasp at the air to catch hold of life as one flails, falling over a precipice? Fear and frustration must be overwhelming in those final seconds.

Today we speak of finding "closure" when a loved one goes missing. Even if it takes years, we are driven to find out exactly what happened to someone when they fail to come home. It struck me that death is a direct parallel. I was the distraught relative. I wanted closure in the matter. June 20, 1969 dawned and my father was still missing.

I was just beginning to realize that no one was going looking for him.

Attempting to lay hold of some answers, I signed up for an elective in school called "World Religions." The class used the textbook *The Religions of Man,* (88) and I remember being surprised that in the study of Comparative Religions, there is not a lot that is comparable. The class gave students an overview of Hinduism, Buddhism, Christianity, Islam and other major world faiths. I was perplexed that – despite the Catholic Church's "ecumenical" teachings – that all faithful people just worship the same God differently – the religions of the world bear almost nothing in common. Even their concepts of who – or what – a god is, are sharply conflicting.

The course did not offer satisfying answers, but it did bring lots of questions into focus. Like many Catholic families, we ate exclusively non-meat dishes on Fridays. One Friday I asked my mom if she would explain why it was sinful to eat a hamburger on a Friday. Her answer told me a lot. "Well," she said, after some hesitation, "I suppose not eating meat is a way of honoring Jesus' death, which was on a Friday, of course." She hesitated, "But it's really more of a Catholic Church rule, though."

So, there it was: churches and people can determine the boundaries separating right and wrong? That *is* what my mom was telling me. It left me wondering what a god's role might be in the equation.

I recalled studying for my first communion when it was required that I memorize the catechism answers. I was stunned when I read that the primary purpose of Man was to "enjoy a relationship with God." I remember thinking, "if that is really the goal of living, why the heck isn't anyone at home or school telling me that?" Instead of life's primary purpose, it seemed more like life's best kept secret.

And then there was something that had happened in church on Christmas Eve in 1971. Father David was a younger priest who was speaking that evening, and as he read from the scriptures, it was as if a camera flash lit up the sanctuary. I don't recall exactly what he said, but from somewhere, the moment was highlighted for me in a pyrotechnic way. He was saying something about God *being* love ... and it made such sense. But the family returned home to the usual sea of paper and presents and there was no interest in exploring thoughts like that any further.

It would be a few years later that a school friend and I took a trip down to Louisville, KY. We were between sessions and decided to visit some associates at a radio station there. The program director gave us the nickel tour and sent us to stay the night with one of their drive-time announcers who turned out to be a shirtsleeve theologian. We rolled out our sleeping bags on the floor at the apartment of a disk jockey who went by the air name "John Dooley." It was not long before we discovered that Dooley had another drive, too. Like a lot of broadcast people, he was always "on" – even when he was not on – and that night John took it upon himself to give us – his captive audience – an unsolicited summary of his spiritual perspectives.

He asked my friend and me to consider an existential riddle. "Picture, in your mind, that your life is a house," he began. "You wake up and your house is on fire. Each window represents a different religion," he proposed. "Now, which window would you choose to climb out of?"

The "Dooley Doctrine," as I came to understand it, could be summed up: "Run to whichever religion is most convenient to you because they will all lead you out of danger."

But had Dooley taken the time to study something even as rudimentary as my high school religions book, he would have known that his hypothesis – as reasonable as it sounded – could not possibly be right. The beliefs of the major faiths are so diverse and conflicting that for any one of them to be correct would mean that all the others were not. Either they were all bogus, or only one was valid and the rest were counterfeits.

I put Dooley's doctrine on my mental shelf for future consideration.

In truth, if my friend and I were part of any kind of faith, it was – as Don McLean wrote in his classic song "*American Pie*," (89) – we "believed in rock & roll. Could music save our mortal souls?" Indeed, I did find myself trying to gain stability in the unending musical stream that wound itself around and through my days. I recall conscious confusion about what the meaning of the word "love" even was. In some song lyrics, love was described as totally emotional or totally physical; in others it was the rush of "love at first sight." Some writers spoke of love as a singular devotion to the point of obsession – an experience so exclusive that a broken heart would break one's own soul. Others advised to just "love the one you are with" and move on. Some songs asserted that true love was naïve and unobtainable. Alas, the ever-present rock & roll in which we were so enmeshed only served to tangle the most important of questions.

A bright sun rose over northern Kentucky. My friend and I headed back home. And much like our road trip, spiritually speaking I was headed right back around to where I started.

Faith in the Science

My interest in the interstellar arrived one day in a small box. It was a gift – a solar system mobile that my parents bought for me when I was in kindergarten. They hung it from the ceiling in my bedroom. You can picture it: the classic design with a yellow molded-plastic sun in the middle. Various planets hung on strings from arms that suspended them at their respective distances. If you walked underneath, the breeze you created would encourage the planets to silently "orbit." Actually, they only kind of bounced around, but it gave you the idea when you were five.

You see, other than my mother's insistence on us attending the mass each Sunday, we really were a very secular family. There was no discussion of things intangible. Dad was an electronics engineer at the Air Force Base. The Mercury, Gemini and Apollo space missions were the news of the day. Our underlying faith was in the trustworthiness of science.

My parents didn't enumerate my personal mission in detail, but I knew it anyway: it was to appear very smart when it came to space. They were only too happy to give me the tools I needed to make that happen. By ten I had been gifted a 4" telescope and spent many a summer evening in the back yard picking off planets when I was allowed to stay up late. I recall people coming around and squinting at the object I had in my sights. I am sure my father was proud that his boy was exhibiting such stellar intelligence with the neighbors looking on.

By twelve I had discovered Sky and Telescope Magazine [90] where I learned that there were a lot more objects within range of a small telescope than the moon and the planets. Elusive nebulae, star clusters and galaxies were like gems tucked among the constellations. In a small instrument, they appear as fuzzy "cotton balls." Such faint targets require a very dark sky to find – like 3 a.m. dark. Locating those deep-sky treasures became my personal challenge.

It was through this late-night sampling of the skies that I was first ushered outside my family's envelope. I would slip out with my telescope to experience the luscious brilliance and stillness of the heavens in the early morning. I saw first-hand the expectant silence and indescribable peace that lie just above the civilized world before a new day is born. Then, about 90 minutes before dawn, the sun's first rays would strike the outer atmosphere, and in an instant, the session was over. The stars, which minutes before had appeared as jewels, suddenly were immersed in the gray fog of dawn that engulfed all but the most brilliant.

I would creep back to bed.

Crawling out a few hours later, I would encounter a world that was already fully engaged. The sun was hanging high in the east. But there, with the light of morning streaking across, lay my notebook displaying the evidence of the past night's adventures. To a banquet of cold cereal, I would celebrate the conquests that were luminous enough to punch thru the Southern Ohio haze.

Those overnight explorations took me outside not only my family's bubble, but also outside the province of regular people. You see, I could not boast that I had identified "The Andromeda Galaxy – M31" the night before. Even if folks knew what a galaxy was, it was not believable that a teenager could see one from his back yard. Even a comment about a newsworthy event like an eclipse or a bright comet would draw darting eyes and polite comments. Despite its prominent placement in the news, space and astronomy were not real to most people, and if you presented that you were even kind of interested in them, suddenly you weren't either. Yet, the nighttime reality that surrounds us fascinated me all the more, perhaps, because so many were so oblivious.

Disillusionment with religion was something I took in stride because our family considered "church" an elective. Science, however, was core curriculum. My father had believed that science was based on facts, not fables, and as a result, my gaining an understanding of science was a requirement. Science had been my dad's religion, so it was through that lens – and through astronomy in particular – that I would try to get a bead on the elusive thing called existence.

I took all the math and science our school system offered: Algebra One and Algebra Two were followed by Geometry, Trigonometry and Calculus. I weathered Biology, Chemistry and Physics 1 – and almost – Physics 2. Through it all, I learned that, at best, I had an average head for science and math. [I had little in common with Einstein, except my wild hair on some days.] Some of the kids in my classes left me standing in the dust, intellectually, and they seemed to do so with such ease while I busted my academic hump. What horsepower they had under their hoods! But even that did not deter me from my want to study the skies.

What did, was my first visit to a school in Indiana. I had landed a meeting with the chairman of the astronomy department – a Professor Francis J Edmonds. He was a stately gent who bore every resemblance to a dusty painting I recalled of the astronomer Galileo, circa 1610 A.D.: white beard; thin, curling mustache. He was the kind of man who probably got roped into to playing Santa each year when the holidays rolled around. Like the painting, he too looked pretty dusty.

The professor cocked his head and eyeballed me from across his wooden desk. He cautioned that only three students in every freshman class of 450 made it through the nine years of science and astrophysics to gain their PHD's. No PHD meant no astronomy job. In short, my odds were awful. Mr. Edmonds made that crystal clear.

Following that meeting, quite perceptibly, my ship began a change of course. It came to a head one day when my high school physics teacher asked me to stay after school. He sat me down and asked me pointedly, "You have taken all the advanced math and science a junior can take here at East Kettering High. You have come all this way and, now that you're a senior, you're not going to take Physics 2?"

I responded in the affirmative. I had just decided not to, I told him.

"That's because...?" he queried.

"That's because I won't need it for what I am going to be doing," I offered.

He asked if I had changed my mind about being an astronomy major in college.

"I have," I said, mustering some confidence in the face of his skepticism. "I have decided that – well – I am not sure what I am going to do, exactly. But I have decided that it has got to be something that's related more directly to people."

"To people?" he said, turning his head to the side a little. I could tell he wasn't buying it. "OK, I just wanted to know if I was missing something." He looked down at something on his desk and paused. "You have always been all about astronomy, Bill, and this seems like a drastic change for you, and rather late in the game, I must say."

"Well, yes, it probably does seem that way," I struggled. "And it does seem kind of sudden, I suppose, but it's been building up for some time."

"Building up?" he questioned, looking up at me. "And yet you are not sure 'exactly' what you want to do?" [He made sure to say the word "exactly" about twice as loud as he needed to.]

I could tell that my comments were not lining up for him and they barely did for me either. He was a man of science and education, I got that. But I also had the impression that he either didn't believe me, or that my not taking his class was an affront to him. Maybe it was a bit of both? What I knew for sure was that I lacked the chops to be his kind of scientist.

As I walked out of his classroom, the unsettled feelings followed. I had let down a teacher who had believed in me. I had squandered his support.

My shoes echoed in the hallway.

But even though I really couldn't give my teacher an adequate answer, I *was* telling him the truth: there was this growing sense that my science heading was wrong-headed. I was starting to recognize that despite my love of the heavens, I was way more interested in people than planets. I had big questions about life and it seemed that the answers were not going to be found in the science lab. But why the sudden turn?

I realized as I was writing this that I really didn't have a good way to explain the change to you, either. Then one night I turned on a late movie and there it was: staring back at me in living color was the red, unblinking eye of "HAL," the "HAL9000 Computer" from the movie *2001, A Space Odyssey.* (91) As I rewatched the 1968 sci-fi classic, I discovered that this movie contained the ideal metaphor to explain my course correction.

The film's plotline: a spacecraft piloted by Earth astronauts was on a multi-year discovery mission to reach one of Jupiter's moons. The astronauts were not aware of the mission's full scope, but it was known in detail to the craft's onboard computer HAL, who was charged with its accomplishment. The purpose of the mission was existential: the continuity of the human race.

The screenplay joined the crew as they went about their daily routines. All was going well until one astronaut made an inconsequential calculation error that was dutifully detected by the computer. It was not long before HAL quietly set out on his own mission – a ruthless quest to eliminate the spacecraft's astronauts, one by one. It first terminated the ones in suspended animation who had not yet been awakened. Then, while on a space-walk, it tried to deny the one surviving astronaut re-entry into the ship. Why had the ship's computer gone postal? It was only logical. HAL had discovered that astronauts made mistakes. Being human, they were not 100% reliable. Therefore, their potential for error jeopardized the mission – a mission that the computer alone understood, could not be allowed to fail. HAL concluded that mission success could best be ensured by the elimination of the spacecraft's most unreliable elements – the humans.

Indeed, it had been building up. Science had been our family religion. I came by it honestly. It had been my trusted ally, theoretically devoid of the fables that plagued churchy religion. I had always believed that science represented facts and objective truth. So, I was not asking of science anything beyond what one would expect from any other system of faith. I just wanted to know how the world began. How will it end? How did I get here? Where am I going?

That is all I wanted to know.

I had been trusting science with a discovery mission that could not be allowed to fail. But I was about to find that ensuring success meant that something would have to go.

A Cosmic Conundrum

Now, the odds are that you are not a scientist. We probably have that in common because I did not become one either. Yes, it was partly because of that dusty professor, but it was more because of the discoveries of a man named Edwin Hubble. You probably recognize his name from the world-famous Hubble Space Telescope that was named for him. But what you may not know is that his amazing discoveries about the universe, instead of answering the age-old questions about our origins, instead, multiplied the questions.

Hubble was arguably the best-known astronomer of the 20th Century, newly hired in 1919 to manage the Mount Wilson Observatory near Pasadena, California. That world-class observatory had just installed the largest telescope on the planet, sporting a mirror that spanned more than 8 feet. Astronomers reasoned that if they could see farther, using a mirror of massive magnification, perhaps they could get a better focus on the elusive structure of the universe.

Elusive? Believe it or not, as recently as 1919, scientists had no clear idea of what was really "out there." Oh, they knew that there was this galaxy that our solar system is in called the Milky Way. It had been visible, even to the ancients, viewed edge-on as a bright stripe in the night sky. But what about the thousands of undefined objects that peppered the cosmos in between the stars? What exactly were those diffuse islands of light that defied resolution? Were they big or small? How far away were they? As recently as a hundred years ago, astronomers still were very fuzzy on what those objects were.

There was also a thorny problem that had plagued scientists since the days of Galileo: even if they could peer farther into space, they had no yardstick to measure how far away things were. Without knowing the distances, they could not map the universe and figure out its structure. To calculate those distances, they just needed to know how bright individual stars actually are, because if you know that, you can easily calculate how far away they are.

If you are driving at night and you see headlights approaching, you know instinctively that the dim ones are farther away and the brighter ones are closer. You know that because you assume rightly that all headlights are approximately the same "actual brightness" so their distances can be ranked based on how bright they appear. But imagine if the intrinsic brightness of headlights varied widely like stars do. You would not be at all sure which ones were close by and which ones were far away.

But just before Hubble showed up at Mount Wilson, scientists discovered a way to smoke out that "actual brightness" mystery. At last, they had a celestial yardstick. They also learned a way to find out how fast objects are moving towards [or away from] the Earth. Those new tools, plus the increased viewing power of the world's largest telescope, would allow Edwin Hubble to become the first person to actually "see" the structure of our universe.

What he observed gave a new meaning to the word "astonishing."

First, using the new measuring methods, he was able to determine the actual distance across our Milky Way Galaxy. He found that it is a shockingly large distance, beyond anyone's ability to imagine. How far? Hubble calculated that our galaxy is so huge that travelling at the speed of light – which is more than ten *million* miles an hour – it would take a person more than 200,000 years to cross it. [Our galaxy is, therefore, said to be 200,000 "light years" in diameter, with a "light year" being a distance – the distance light travels in a year.]

Back when Edwin Hubble was in school, he had been taught that our galaxy was the only galaxy that there was. The theory, at the time, was that everything you could see in the night sky was contained within the Milky Way. Those "fuzzy objects" among the stars were assumed to be big clusters of stars within our galaxy that were just too far away to be made out clearly.

But Hubble's research proved otherwise. Some of those fuzzy objects were so far away they weren't even in our galaxy at all. [That explained their "fuzziness."] In reality, they were totally separate galaxies, all their own. For example, he found that the "Andromeda Nebula," as it was then called, was really 2.5 million light years *outside* our galaxy.

And Andromeda was merely the closest galaxy. There appeared to be countless more, scattered throughout the heavens in every direction. When the telescope that bears Hubble's name was launched into space, it would allow us to see the universe from a vantage point outside our atmosphere. It proved what astronomers of Hubble's era had suspected – the number of galaxy worlds that we can detect – even just from our limited vantage point – is as un-countable as the stars themselves. There are so many that no one could, even if they worked their entire life, just number them all. The observable universe, which was once suspected to be vast, was now understood to be infinitely vast. And our Milky Way home is but a modest-sized island within the infinite.

If that wasn't enough, Hubble discovered something even more unexpected: all of the galaxies were moving *away* from ours. Not one of them was approaching. This led him to the hypothesis that the universe is expanding outward from a central point. The only way that could be, Hubble believed, would be if all the universe's matter and energy had all been in one location in the beginning of time and had exploded outward – in a "big bang." All the material would have been flung from this central point in all directions, and therefore, from the perspective of any one piece, every other part continued to recede indefinitely. Since space is a vacuum, there is nothing to stop things from expanding. Working backward, based on how long it would have taken the universe to increase to its present size, Hubble calculated that it would have started expanding almost 14 billion years ago.

Now hold that thought for just a second. Remember my solar system mobile? It was held together by string. But what holds the real solar system – and the entire universe - together is not string, of course, but... gravity – a force that comes from matter itself. Yes, the "stuff" of the universe actually creates gravity by its very presence. And the more stuff there is packed into a given area, the more gravity there is to pull inward. That is why you and I are not floating in space right now. Earth is massive and so its gravity keeps pulling us down toward its center. Gravity can tug not only on you and me, but also on every person and skyscraper, the entire atmosphere of air, and even an orbiting body the size of the moon at a distance of 250,000 miles. And the larger planets like Jupiter have more gravity than the smaller ones. This was nothing new. It was known long before Hubble that the larger a body, the more gravitational grip it has.

So, some scientists had begun to suspect that there could be bodies so massive that they would have enough gravity to pull everything into their grasp, like cosmic vacuum cleaners. They would suck up every bit of nearby matter – and even energy – like light itself. They were dubbed "black holes." Their gravity would be so intense that nothing that fell in – by definition – could ever fall back out. So, a black hole could only grow larger and more massive and powerful as it gobbled up surrounding materials and even whole stars, planets and galaxies. Einstein predicted that black holes must exist, and that even light itself would be subject to their gravitational pull. [Hence, a *black* hole.]

Today, scientists know that black holes are very common. Most galaxies like ours have massive black holes in their centers, some of which are so densely packed that the contents are equivalent to millions of stars the mass of our sun. The amount of gravitational pull of such a body would be enormous.

Perhaps already you see the contradiction.

In the beginning, if the whole universe had really been contained in one place, as is accepted in many quarters, by definition, all that material would have possessed so much gravity that nothing could have escaped. The entire universe would have been comprised of the most massive black hole imaginable, containing all the world's matter and energy packed tightly into one space. By astrophysicists' own definition, it could never have exploded. Nothing within it could ever have escaped. And the materials inside would have been squashed into a formless substance, not conventional "elements" as we know them.

So, how do astronomers explain how a black hole, which by its very definition could never explode, would have exploded with a "bang" so big that it filled the entire universe with its contents? There is only one way it could have done so, according to scientists, and that is if the very physical laws that govern our universe were not in play for the first few moments of the explosion. The universal laws of physics would need to have been suspended for a fraction of a second to allow the explosion to occur. Only then, after that fraction of a second elapsed and the expansion was underway, would the known properties of physics have come into being.

Really? So, the raw materials of the universe actually would have existed *before* the laws of physics that now govern them? Then something unknown caused it all to explode? Then, beginning a fraction of a second after the explosion, only then did the physical laws come into being?

Does that sound preposterous?

I read these theories as a young man and concluded that scientists had painted themselves into a cosmological corner and were desperately looking for a way out.

Random Acts

If you think I must have become a "science-denier" in my youth, nothing could be farther from the truth. Though theories and hypotheses can be way off target, the scientific method and the data and findings that it produces are sound applications and always will be. Obviously, good science can generate accurate results and has yielded many, many great discoveries. Where things begin to break down, in my estimation, is when the mission begins to weight the data in order to arrive at a predetermined conclusion. That is, when good science data are interpreted to *persuade* rather than explain. Examples from daily life – especially as evidenced in marketing and advertising – are plentiful. How many new product ads encourage us to make a purchase based on a "scientific breakthrough?" And who can forget "follow the science," the slogan we remember from the pandemics? That phrase became associated with the contradictory advisories that government officials issued, some of which turned out to be correct.

As this relates to the universe's origin, I came to believe that scientists had begun asking the wrong questions. Notice the not-so-subtle difference between, "how can we discover the origin of the universe" and "how can we discover *how the Big Bang Theory* explains the origin of the universe?" Asking the second question is how you explain scientists putting forth hypotheses that require the laws of physics to take a dive for a few seconds while the universe gets underway.

But even more disappointing to me is the continued appeal to "infinitely-occurring random activity" to explain the universe and its elaborate order. Give that some thought for a moment. Where, in life, does increased disorder lead to increased order? Certainly not in the world we all live in right now. And yet, theorists repeatedly have proposed this as the basis for our very existence. Can it be true that century after century of directionless, random activity will eventually result in an elaborately-ordered universe: the elements, the stars and planets, living things and even intelligent life? Yet this subtle tenet is woven throughout modern scientific thought. I encountered it so frequently that I made up an acronym for it: "R.A.N.D.O.M." – the 'Random Accidents Naturally Develop Order' Mechanism.

Below are a few well known examples:

How did the materials in our universe originate? The Big Bang explosion, that theory says, resulted in a rapidly expanding mash of matter, energy, space and time that had absolutely no form of any kind. Now, you and I cannot imagine "no form" because we have always known only things that have "form." So, "no form" is the very absence of what we can imagine. That is what scientists say would have resulted from the "Bang" – an unimaginable substance slinging in all directions.

The theory says that this original substance would have been blasted throughout space – beginning at the point of the blast - at random. That meant that sometimes there were random collisions among materials, and those collisions eventually could have resulted in uneven densities – clumps of material – which, like snowballs, attracted more material by their gravity and got even bigger. Given a virtually infinite amount of time, they say that the contracting materials would have condensed into the 188 elements and begun to rotate in disks and became the stars, the planets and the galaxies.

So, the theory goes, the Earth would have started out as one of those rotating clumps of material, revolving around a larger one that had more mass so that its intense gravity caused it to compress and ignite. That would have become our sun.

In this way, R.A.N.D.O.M. activity is proposed as the sole engine responsible for the formation of the Earth and all of the heavenly bodies. 99

How did life come about? All life on our planet consists of microscopic cells. So, for life to have begun, one first must propose how a living, functioning cell gets constructed from inanimate materials like rock, sea water, air, etc. Scientists admit that the raw materials to make the first cell would have all been inanimate – not alive.

But the first cell would have needed a cellular wall that could be maintained by the organism and knew how to grow on its own. It would have needed the ability to process other molecules for food. It would have needed to divide itself so it could reproduce and make more cells of its kind.

But do some research. You will discover, as I did, that scientists do not know how an intricate organism such as a living cell could form on its own from lifeless materials. There have been experiments in the past that purported to show random activity like lightning bolts striking seawater a virtually infinite number of times - which was the going theory when I was in school - or the undersea action between rock and warm seawater (called the "hydrothermal vent theory") across millions of years – could have created *conditions* in which substances such as amino acids could have developed. Amino acids are a key part of living cells, but they are not alive. And experiments with electricity and salt water have repeatedly failed to lend any support to such theories. In fact, you will not find any specific hypotheses that propose how the first cell wall, the first organic molecule (DNA) and the first highly-intricate cellular programming came about – all of which would have been required *simultaneously* for the first cell to be constructed, to obtain energy, to grow and to reproduce – to become alive. Obviously, if there is no explanation for the existence of the first cell, there is no explanation for the existence of life.

If there truly is a "missing link," it is right there - at the intersection of the inanimate and the animate.

Intelligent Life It has long been known that when living cells reproduce and divide, sometimes they will do so imperfectly, resulting in a mutation – a random replication error. Generally, mutations produce life forms that are imperfect: a rabbit that has one leg with an improperly formed joint cannot run as fast as others of his species and gets gobbled up by a predator. Animals with mutations generally would be compromised and would not have the opportunity to produce as many offspring. But the theory proposes that sometimes, quite by accident, a random mutation could produce a superior animal – one which is more apt to survive precisely because of the mutation. There would eventually be more of that type of animal, and therefore, more offspring. Eventually, these "better animals" would become the dominant type because their random mutation made them more plentiful. This "Survival of the Fittest" hypothesis, which Charles Darwin dubbed "Natural Selection,"(92) is proposed to be the driver that propelled the offspring of the first cell through a virtually infinite number of random mutations until the result was intelligent life – human beings.

But there are obvious problems with this hypothesis. To name just three:

The Earth's fossil records do not indicate the kind of gradual changes in animal and plant forms which would have been the result of Natural Selection. Instead, fossils show sudden appearances of radically new life forms and sometimes the abrupt disappearances of others.

Natural Selection also doesn't explain the advent of new, highly-advanced animals or plants. For example, the very first flying animal would have had to have been immediately preceded by a parent animal that did not fly. Natural Selection does not explain how all of the required body changes – such as a pair of opposing, aerodynamic wings, light-weight hollow bones, a feathered skin covering and a brain that instinctively knew how to synchronize it all - could have occurred on the very same animal simultaneously. If a random mutation had caused even one of those changes to occur, the animal would not have had any flight ability and the change would not have been retained, per the theory.

But perhaps the most glaring problem is that the Natural Selection theory fails to explain how there are 9 *million* unique species of plants and animals that live on our planet simultaneously. For instance, there are more than 11,000 different kinds of birds alone. And it is unexplained why there would be thousands of different species of animals and plants that cohabit the very same acre of land at a given moment, all of them competing for the very same energy resources. Instead, the Natural Selection theory would predict that there should be a very short list of highly successful animals and plants - not a virtually inexhaustible one.

The myriad difficulties with evolutionary theories are well documented elsewhere. I mention Darwin and Natural Selection here solely to illustrate that this hypothesis too rests on R.A.N.D.O.M. for support.

Please understand, I did not expect that scientists would have rock-solid answers that explained every development process that has occurred in the history of the universe. But as a young man, I was disillusioned when I observed that so many of the accepted explanations about our origins depended not upon actual evidence, but upon the roll of the dice. The repeated reliance on chance as the "go-to" explanation for our existence led me to be very skeptical.

And regarding the reliance on chance: it is well known in statistics that if you roll a pair of dice an infinite number of times, every possible result will eventually occur. You will roll a "6" a thousand times in a row. How many times will that happen? An infinite number of times, since you are taking an infinite number of rolls. The point is, anytime the word "infinite" is part of the equation, very unlikely results are guaranteed. Is that why scientific theories that rely on infinity are proposed to explain outcomes as unlikely as intelligent life?

But the root problem with applying "infinity" to theories about our origins is that the universe is not infinitely old. Taking science at its word, it is about 13.8 billion years old. Though 13.8 billion years sounds like infinity, it is not. 13.8 billion is a finite number, and it is not really that large. For example, the United States spends 13.8 billion dollars every few days. So, it is disingenuous for scientists to suggest that "because the universe is infinitely old, anything can happen." They themselves tell us that universe is *not* infinitely old, and therefore specific outcomes have specific chances of occurring.

Following that logic, it forces the question: is 13.8 billion years long enough to permit a sufficient number of random things to happen so that formless material flying about in space would transform itself – on its own – into an orderly universe of stars and planets, 9 million unique life forms and billions of intelligent human beings? Because "chance" has been introduced by some scientists as the sole mechanism responsible, it is fair that we would ask them to state just what those chances actually are.

Spoiler alert: you will not find an estimate of the odds of formless matter becoming intelligent life published anywhere.

The Alien Absence

If R.A.N.D.O.M. is right – if disorder really becomes orderly - all by itself – then alien beings must be virtually everywhere. Intelligent creatures should be in various states of evolutionary development on countless planets throughout the universe, right this minute. This is what most scientists of our generation firmly believe. Not only are we "not alone," they say the galaxies must be fairly teeming with alien lifeforms in all stages of evolving. And many of those alien civilizations should be millions of years more advanced than we are, giving them a massive head start on developing the technologies to communicate across the expanse of space.

This is not a new notion at all. The idea that alien cultures could be monitoring our electronic transmissions, or sending transmissions to us, originated in the 19th century at the time radio was invented. In fact, the best-known radio pioneer – Guglielmo Marconi – speculated that "wireless" signals could be used to contact intelligent beings on other planets. And, in a way, he was proven right when radio was the very method used for communication with the first lunar astronauts.

But by the 1950's a more considered hypothesis was being advanced. Astronomer Frank Drake calculated the approximate number of stars in our own galaxy and produced conservative estimates of how many, potentially, could have Earth-like planets circling them. Then he went on to guestimate how many of those planets might be developed enough to host intelligent life with advanced technical abilities. Astronomer Carl Sagan of PBS's *Cosmos* (93) relied on Drake's numbers when he asserted on national television that there are millions of civilizations in our galaxy alone that have already developed technology to our level, and beyond. Sagan went on to predict that since there are an uncountable number of galaxies, it is just a matter of time before we receive a communication from an intelligent being from another world.

With the encouragement of these and other prominent scientists, another kind of space program you've probably never heard about was launched. This program was named "S.E.T.I. – the Search for Extraterrestrial Intelligence." Yes, it does sound like a hippie cult from the '60s, but in fact, it was an elaborate effort to detect alien communications on an international scale. The S.E.T.I. programs in the U.S. had mainstream funding from big players like NASA, the National Science Foundation, Ohio State University and spokespersons like Sagan and physicist Stephen Hawking. The effort splintered over the years and changed focus a number of times, but the premise has remained: "Millions of intelligent cultures that surround us in space are very likely trying to communicate. They will choose common technologies like radio waves or lasers to create beacons and beams which can be transmitted efficiently among inhabited worlds. The first communications would likely be universal mathematical axioms so the receiving culture would immediately identify the sender as intelligent."

One of the first major listening efforts was Ohio State University's massive 20-acre radio telescope near Delaware, Ohio. Built in the 1960s and dubbed the "Big Ear," the device functioned for more 35 years, scanning the night sky for extraterrestrial signals. Similar efforts were begun at the giant Aricibo Radio Telescope in Puerto Rico and the National Radio Astronomy Telescope in West Virginia. A Harvard-Smithsonian device in Massachusetts joined in the effort as well. All were designed to systematically survey space, monitoring a specific set of likely communication channels. Later, with advances in technology, better receivers were designed which could monitor billions of frequencies simultaneously, making the results impressively comprehensive.

The S.E.T.I. strategies were simple: operators aimed the listening antennas methodically at different slices of sky every night as the Earth proceeded around its solar orbit. As a result, virtually every square inch of space would be surveyed multiple times. And after the US launched the Hubble telescope and astronomers were better able to detect planets orbiting nearby stars, listening efforts were targeted toward the very parts of the sky known to contain those planets.

Though S.E.T.I. efforts are no longer funded by the likes of NASA or the National Science Foundation, they continue today in limited ways, made possible through private donations.

In Table 1, below, I have compiled what I believe to be a complete listing of all of the verified extraterrestrial transmissions detected in the last 60+ years through the "Search for Extraterrestrial Intelligence" Programs. (94)

Table 1 – S.E.T.I. Extraterrestrial Contacts

Year	Alien Signals Detected
1957	None
1958	None
1959	None
1960	None
1961	None
1962	None
1963	None
1964	None
1965	None
1966	None
1967	None
1968	None
1969	None
1970	None
1971	None
1972	None
1973	None
1974	None
1975	None
1976	None
1977	None
1978	None
1979	None
1980	None
1981	None
1982	None
1983	None
1984	None
1985	None

1986 None
1987 None
1988 None
1989 None
1990 None
1991 None
1992 None
1993 None
1994 None
1995 None
1996 None
1997 None
1998 None
1999 None
2000 None
2001 None
2002 None
2003 None
2004 None
2005 None
2006 None
2007 None
2008 None
2009 None
2010 None
2011 None
2012 None
2013 None
2014 None
2015 None
2016 None
2017 None
2018 None
2019 None
2020 None
2021 None
2022 None
2023 None
2024 None

Table 1 illustrates, better than words, why funding dwindled for S.E.T.I Programs. Remember, a broad consensus of renowned scientists had estimated that there are *at least 100 million* intelligent civilizations in our galaxy alone, and, as we know, there are a virtually infinite number of galaxies. How can it be that we are receiving from these billions and billions of hypothetical alien civilizations... absolutely nothing? After more than 60 years of surveying every part of the sky, there are zero detected signals.

Silence.

Meanwhile, the waves from Earth's daily telecommunications radiate into space continuously like ripples on a 3-D pond. They move outward daily in an ever-expanding bubble at the speed of light. Earth has been leaking electronic waves from radio stations for more than 100 years now. The result: within 100 light years of our planet, we have quite unintentionally announced our presence to all cultures capable of receiving and decoding simple radio signals. Yet we are not receiving any responses from our intelligent interstellar neighbors?

Silence.

Recently, we even began a program named "Breakthrough Listen" that utilized the most advanced 21st Century monitoring technology to survey 1,327 of the closest stars for precisely the types of unintentional signals that our own civilization produces.

Same result: silence.

Dead silence from space strongly suggests that R.A.N.D.O.M. is dead wrong. If disorder naturally results in order, by scientists' own estimation, space would be a very, very noisy place.

Indeed, the *2001 Space Odyssey* movie does present the ideal analogy to explain my abrupt change of course: since science's mission to discover our origins has been based on random odds and chance, the death of commonsense will absolutely be required to ensure mission success.

The Grand Point of Sand

I cracked open one eye to survey paradise: a jet-blue-sky and a hot sun were arched over my favorite beach; a breeze was wafting a Coppertone aroma among the sun worshipers, all of whom were my very best friends. For me, it was a personal "mountain-top moment." Life was good. I was working in the career I had intended and had been promoted to the position I had aspired to. For the first time, I owned a new car and had enough vacation time and money banked to make the trip possible. I had even cued up some tunes from my very favorite band. Of course, a cooler was at hand should I need refreshment. It had taken a good part of a year for me to pull all the right people and all the right paraphernalia into this perfect space. I knew it could not help but be good. As I closed my eyes again, I congratulated myself that I had accomplished it.

But surveying it all in my mind, I had to admit that I was nothing short of miserable.

The problem definitely wasn't the surroundings. This beach wasn't just any beach. Jutting westward into Lake Michigan, on the right day, Big Sable Point has arguably the most agreeable conditions to be found anywhere. Mounds of soft, "sugar sand" stretch to the horizon. Its rounded amber, black and cinnamon-tinted grains are whipped by prevailing winds into imposing dunes that stand as sentries at its portals. Named by the Frenchmen who first saw it, the place name means "Great Point of Sand." Indeed "Grande Pointe Au Sable" is the largest freshwater sand dune formation anywhere on Planet Earth.

It was at this very beach that my family enjoyed its last vacation when I was a boy. It was to this beach that I would return to search for something to grab onto. I was more than a little disillusioned with the life guidance I had received from religion and science. So, I had freed myself of such entanglements so I could focus on my personal development.

Yet there was something about the place that was unsettling. I had kept a map of its trails during my teen years and even recreated them in the muddy Ohio woods. Mine were named after the Point's sandy, sun-dotted originals. But something would draw me back, and once I could drive, I proposed a return. Mom was able to hold me off until the summer I was seventeen.

My companion in the two-week adventure was family friend George, my elder by two years. To him, my mother had inexplicably entrusted our metallic-blue 1967 Le Sabre Buick, our family's camping equipment, and me. Though we had intended to camp at a number of places, once we pitched our tent at the state park on Big Sable Point, we never considered going anywhere else. My nineteen-year-old friend had always had that "mother hen" personality, and though I never asked him, I was certain that he had been deputized by my mom to keep me corralled.

George's family had lost his dad to divorce about the same time we lost ours, so our family remnants ran with each other for a time. His mom and mine leaned on each other in those first days and he and I had a lot to compare nightly around the campfire. We would stay up until all hours, agreeing not to crawl into our sleeping bags until every glowing coal was punched out by our pokers. That was often just before sun-up.

What I did not anticipate on that trip was what awaited me – not in the campground – but at the lakeshore.

The drive from Southern Ohio to Big Sable Point takes a half-day to accomplish, even at 70 mph. Modern "superhighways" I-75 and I-96 had been completed by the 1970's, allowing us to make the trip in 8 hours. To beat the traffic, we started north before dawn and arrived at our campsite just in time for lunch. Unfortunately, the peanut butter had been packed in the car's back window and it was easier to pour the goo out of the jar than to knife it onto the bread. We couldn't have cared less.

After we set up our tiny canvas tent, George suggested we go to see the lakeshore which was so close we could hear the roar of the surf. He and I followed the ascending sound into the woods and out again; up a sandy incline that ended atop a steep dune where a solitary bench was perched.

The scene was just as I recalled from boyhood: the wind-swept shoreline was a full 180 degrees of stunning. The water was painted every imaginable hue of green and blue, topped by breakers whipped up like vanilla icing. The 180 degrees behind was a field of amber dunes crested with scruffs of marram grass. A gallery of old growth pine waved its boughs wildly as if inviting us to wave back.

George turned to go back to camp but I couldn't just yet. "I'm going to the store," he advised as he started down the sandy path. "See you back at camp?"

"Yep," I affirmed, as his tee shirt, red hair and vented ball cap disappeared from view. I would follow him in a few minutes, but first I needed some time to take it all in. There was the sun, the sand and the surf I so vividly recalled. It was still there, as if it had been waiting for me. But there was something more. Something at the shore awaited discovery. I would have two weeks to figure out what it was.

When I got back to the campsite there was a note fluttering on the picnic table, pinned under a camping lantern. George had departed. I saw the opportunity to grab a few minutes of shut-eye on a newly unrolled sleeping bag. I awoke to the sound of him returning, settling into one of the camp chairs.

The Earth had turned while I was asleep and sunshine had replaced the shade across the tent's west side. Shadows of waving limbs raked across the green canvas to the music of the wind. As I came to, I realized that the tent interior had become a sauna. But more surprising was the vivid dream that fought to continue as my eyes opened – a dream so real that I considered going back to sleep to see what would next happen.

I was on the beach, digging furiously with my hands to unearth a treasure. It was in vain. I looked around in frustration. I was alone. The beach was vast and I was exhausted. I might dig forever. I was just beginning to excavate in a new spot when I sensed that someone was behind me. I could only see the slender arm of a girl which reached into the hole where I was digging and pulled up a fist-full of wet sand. She held it before my eyes. Her fist was so tightly clinched that it shook. Then, in an instant, it relaxed.

Her hand began to open, one finger at a time. The damp sand had been compressed into an irregular shape that showed clearly the curves where her fingers and palm had mashed it into concave scallops. As the sunlight shone on it, its surface began to dry. Cracks appeared. Then the sand began to fall away, one grain at a time.

Peering through widening crevices, I could see a blue glow illuminating the crumbling mass from within. As more and more sand dusted away, I was able to make out the source - a tiny gem – intensely bright. Its color was a deep sky-blue with faint white lines that seemed to dance on its surface. I started to ask what it was, but as I did, I awoke.

The reality of the musty tent air gradually supplanted the images. All the while, through the tent's screen door I could see outside to a world where a fierce wind blew and spackles of sunlight danced across a bronze carpet of pine-straw. The tent's portal looked as if it were a painting, but one in which the framed image was alive. A moment later, I would pull up the zipper and step into that image, and as I did, I was oddly aware that a transition had begun. I looked around to see George intently polishing his new leather boots. The climb up the dune had dulled their shine and he was already making ready for our next excursion.

Daily hikes with George were more like marches, always with him in the lead. On nice days our return to camp was followed by a visit to the beach. But it was not long before George's fair skin turned to red. Then it became way too red for more sun. Beach visits became solo excursions for me while he stayed in the shade. So, I sought out a prime spot with a view of the shore. My perch ended up being a lone dune woven with sand cherry roots – a high point that the relentless winds had not overcome. That would become my fortress, my lookout, and without George on sunny afternoons, my place to process.

After dinner and sundown, George and I would often sneak out of the campground for stargazing and meteor watching. Ascending a dune and plopping down into sand still warm from the day, the angle was perfect for peering up at the sky. The Milky Way manifested as a stripe of gems strewn across black velvet. Silence prevailed, except for the occasional tumble of a wave.

One day from my perch I spied a stylishly dressed man who arrived at the beach parking lot in a very expensive-looking sports car. He parked near the entrance at a diagonal so that no one could fail to admire his ride. From what I observed, few missed it. And from my angle of view, it appeared he had parked directly beneath a tiny birch tree that grew on the south edge of the lot. I observed as moms and dads in flip-flops, kids on skateboards, young adults holding hands and older couples paraded past his car as they approached the beach for a day in the sun. Not surprisingly, no one looked up at the little tree, but the juxtaposition of the two was striking.

It triggered a cascade of thought.

Then Saturday, July 22nd dawned and something new began to stir. The hot sun proceeded to climb over the piney canopy like it had most every other day. The beach fog burned off just as the voices of vacationers faded up, drowning the squawks of the lake birds which had owned the shore a few hours before. But on this day, it would be a gusty south breeze that would whip the lake water into walls of foam. The surf had transformed into an aquatic adventure for those who dared dive in.

All at once fresh faces approached, racing from a maroon van into the sea amid shouts and shrieks. There must have been a dozen kids in the entourage. Adults followed with the requisite towels, chairs and inflatable devices. I gathered what courage I could, convincing myself that these strangers needed to meet someone new.

The rollers rose and fell in succeeding swells, requiring one to jump high to keep the head above water. Receding, it gently landed feet back on the sandy slurry that was the lake bottom. In a practiced maneuver, I bobbed among the waves and worked my way toward the group, ending up in their midst as if by accident. With proximity too close to be ignored, I found myself in conversation with one of the guys who appeared to be about my age.

"What a great day to be in the lake," he shouted, popping like a cork above the surf, then disappearing in a splash. I could tell he was more accustomed to the big waves than I.

"Where are you from?" I returned.

Though I was making friendly talk with the boy, my real interest was a girl in the group who had long dark hair. Her name was "Monica" I would soon learn.

I scrambled out of the water with the kids as if I were one of them. On the shore, others were just completing the beach-burial of an unlucky comrade whose head and neck were the only body parts still visible. A young girl with a film camera snapped pictures while Monica and others clapped and giggled. Then the sandman emerged from his underworld looking like a cinnamon doughnut come alive. He darted into the surf to dissolve his coating.

Then it was time for the group to leave for lunch. They departed as quickly as they had arrived. I returned to the campground to eat with George but the site was already so hot that we decided to splurge on a burger in town. On the way back we visited the grocery store and slow-walked the aisles to bask in their luscious air conditioning. Then we stopped to get air in one the Buick's tires. Arriving back at the beach, I assumed my afternoon watch. It was then I again saw the group, this time lounging on the sand by the beach house. Wrapped in fluttering, half-damp towels, they looked to be worn weary by the waves.

I spotted the Monica girl among them so I decided to walk in their direction to see if I could connect. It worked. The boy from the lake hailed me. I grabbed an open spot on the sand that turned out to be right next to some of the younger kids and the adult sponsor. From that vantage point, I was able to confirm that Monica and the sandman were an item. They departed together hand-in-hand for a walk on the beach. Looking around at the kids who remained, at 17 it appeared I was a bit old for their company. As you might imagine, I wasn't planning on spending the afternoon with a chaperone and the chaperoned.

"Where did you go?" piped up a young girl. I recognized her as the one I had seen earlier with the camera.

"We went into town to get some air in one of our tires," I stated flatly. The questioner was not of driving age and I didn't expect any follow up, but I got it.

"Tars?", she said, making light of my southern-sounding speech. "Where so ever are you frum?" she challenged, feigning an exaggerated accent. She looked directly at me with a pair of friendly brown eyes. "Around these parts we say 'tires' with a long 'i'." That got some giggles. Apparently, a number of the kids thought my Ohio accent sounded hayseed.

"Well, I'm from waaay down south in O-hi-o," I countered. "We do and say a *lot* of things different frum you-all." They got quiet, thinking me offended. I continued, "Where I hail from, it's against the law to talk like you have a clothes pin on your nose..."

Oh, they liked that one. Apparently, that was just the kind of cheeky response they appreciated and it was not long before I was getting along with them well. Before long, the sponsor got up to bring things back to the van and with her went most of the kids. Seated by me, the camera girl and her friends remained.

"So, you are not here by yourself?' the girl inquired. "You said '*We*' went to town."

"My friend George and I are camping here," I explained. "He and I drove up from Ohio last week. He has red hair and can't take all the sun."

It was in that exchange, and a few that followed, that I befriended this youngster, who I learned was with a youth group on a weekend camping trip. It was uncanny. Despite her tender age, she seemed to have a natural steadiness and a genuine curiosity about people that came from a place I had never before seen. She didn't seem to be interested in talking about her girlfriends, her boyfriends or the next week's party. Instead, she wanted to talk about life. I thought, "How ironic that no one in Ohio – not even my own family – has taken an interest in discussing topics like that in any depth, yet this 14-year-old kid...." I was unprepared – and yet – I absolutely welcomed the moment. She asked about my family and I ended up telling her about my father's demise.

Her expression changed and she began to speak more intentionally. "Remind me to tell you something," she said after a pause. "And," she continued, "what campsite are you and your friend at? We will come by after dinner and take you over to the campfire sing-along in the Cedars Camp."

George and I were still drying dishes when we spotted the girl and her friends ambling down the sandy road toward our campsite. George looked up at the approaching females, cracked a half-smile and said, under his breath, "Stellar, William. How did you pull this one off?" I only grinned, knowing that I would only have a few moments of glory before he would realize that the girls were all middle-schoolers.

George opted to stay behind while I tagged along on what turned out to be the best evening of the entire trip. The fireside sing began just as dusk was falling. Sparks rose into the starry sky as song after song was raised. I did not recognize a one of them but that was okay. The comradery provided a much-needed counterpoint to what had become a surprisingly solitary excursion.

When the singing was through and it was time to retire, the girl asked me to wait. After darting under a tent flap, she emerged, seemingly empty-handed. Then, she handed me a sheet of torn-off notebook paper with an address scribbled on it. "You *have* to promise! Write me when you get home," she said with an intense look. "And always keep this," she insisted, pressing a small metal object into the palm of my hand. "I know you are going to make it." I looked down and the object in my hand was something she had taken from her necklace.

That is when I turned and left the friendly firelight.

There was no moon that evening and with no flashlight, there was no way to see beneath the dark canopy of hemlock that hung above the foot trail that led back to our campsite. Instead, I would follow the paved roads. The air had turned cold and a lake fog was beginning to settle in. I thought I knew the camp by heart, but it was so dark and so foggy that unless a car was nearby to splinter some light through the trees, it was impossible make out what I was about to walk into.

As I felt my way along, I pondered "Just what did the girl mean when she said 'I know you are going to make it?'" I had assumed she was saying that because I had no flashlight. Or was she referring to something about my father's passing? There was something I was missing. She spoke as if she knew something that I didn't. Frankly, it nagged at me that a middle schooler might think she could see something about me that I could not.

Eventually I found my campsite and was heartened that there was a fire still burning. As I got closer, I spotted George in a camp chair seated next to someone I did not know. They were both poking at the fire. Laughing furiously, they tried to keep a hushed tone as I walked up. As soon as I opened my mouth, they exploded like a pair of school girls.

As long as I stood there, they would not stop snorting and giggling, so I crawled into the tent for the night.

George and the stranger had obviously been smoking and when I asked him about it the next morning, he denied it. Perhaps he didn't want my mother finding out? But the ease with which he pushed aside our friendship was more than disappointing.

Two days later as we packed the Buick to go home, one thing was certain: the trip to Big Sable Point had been a ground-breaker, yet I had no inkling about the misery that lay ahead. I remained convinced that science and religion didn't hold life's answers, but like a starstruck contestant on Wheel of Fortune, I already possessed all the letters I needed to solve the puzzle.

But getting it right - that would mean the game was over.

The Shiny Object

They call it "Monsoon Season" in Indiana – the time each year that begins in November and lasts way longer than it should. Students slog to and from classes wearing wet, oversized backpacks that don't fit under their umbrellas.

The collegiate culture was teaching me that my life should consist of embarking on the right career; meeting the right girl; obtaining the right tangibles. I was heading into a world crowded with billions who were elbowing for those very same things, so I reasoned that if I did not take, I would not have. If I did not have, I would not be happy.

But there is a heavy kind of darkness that wraps around you when you only wrap yourself around yourself. My dark stumble through the campground was becoming my "life analogy" as I wandered among relationships and preoccupations in search of something stable. Light was hard to come by. Friendships were tenuous and temporary. It was obvious how selfish were most people, yet I was unable to see that my focus on me was the source of my very own darkness.

And yet, even during those days, there were some glimmers.

The girl with the camera and I became loyal, letter-writing "pen pals." We conversed regularly by mail as she proceeded through high school and I, through college. I looked forward to receiving her letters in my dorm mailbox, and though I could not have told you how to categorize her, she was always there.

And speaking of the dorm, I often encountered a tall, lanky upperclassman I only would remember as "Gary." He had a sandy mop of fair hair that rode atop two bushy eyebrows. He was exceedingly friendly, and often spoke to me as if we were the oldest of friends. In fact, I barely knew Gary's name, but somehow, he always knew mine. He was part of a religious group of some sort that met in the dorm.

Their meetings took place in the room directly above mine. Monday nights they sang songs and whooped it up. It wasn't heavenly. There was a heat pipe that ran down the wall between our rooms and the sound of their carrying on would reverberate up and down as I was trying to fall asleep. They were annoying to the nth degree, and yet I just could not make myself tell them they were keeping me up. It was partly because I didn't want to be their Mr. Grumps. Plus, is it reasonable to ask anyone to be quiet in a college dorm at 8 pm? [Nobody else on my floor had a job that required them to be up at 3 am.] But it was because Gary and his gang seemed like such good eggs, I kept quiet.

But the biggest glimmer appeared the afternoon of 7th of April in 1977. Trees blossomed on the campus as a warm breeze stirred. Fledgling leaves struggled to spread shade over new shoots of green grass. Just after lunch my dorm room telephone rang. I did not hear it because I was outdoors enjoying a moment in the sun. My roommate sauntered out to let me know that someone "important sounding" was asking to speak to me. The call turned out to be my ticket out of Indiana to my first job at a small radio station on the Mississippi River in Moline, Illinois – one of the "Quad Cities." I looked at the date (4-7-77) and thought, "Now that is one date that I will not soon forget."

As it turned out, I would not be allowed to.

When I called back to Ohio to tell my family the news, I was told that my sister was giving birth to her second child at that very hour. That was noteworthy. In the Wilkins family there had been only one other child born in the previous 20 years – her first, ten years before. It would be another eleven until we would see another.

A few days later I began the drive to Illinois where I hoped to locate an apartment. But as I approached in the dark, I misjudged the exits in the fog and crossed the Mississippi River into Iowa instead. I spent my first night in an Iowa hotel and found my first apartment in Davenport,

Iowa. Largely because of my navigation error, I would end up being an Iowa resident who worked across the river in Illinois.

Right away the apartment needed telephone service, and back in that day, to order a phone you had to physically walk into a phone company office. You couldn't just set up an account on-line because there was no "line" yet to be "on." The internet was still 15 years away. And, of course, you couldn't call to set it up because you didn't have a telephone yet. So, I borrowed a "telephone directory" – a very thick book with very thin pages – to find the address of the nearest Iowa Bell Telephone Company office.

Back then, you were not permitted to ask for a specific phone number. You just received the next one that came up. To learn the number, you looked at the laminated, cardboard circle in the middle of the telephone's rotary dial. The lady handed my telephone to me across the counter in a white plastic bag ...along with a phone book and a bunch of advertising sheets. It was heavy. 1970's-era phones had a solid steel chassis. The bag probably weighed 10 pounds. I looked inside. The phone's hard plastic exterior was egg-yellow. It had a long gray cord with a plug with four brass pins in a square pattern that fit into a wall jack with matching holes.

There, in the middle of the dial was my new number: The prefix was "3-2-2" and the four-digits that followed were "4-7-7-7."

"Really?" I thought. It would be an impossible coincidence that in Indiana, a phone call had made me notice that date, and now in Iowa, it was again the telephone that would focus my attention on that sequence. Plus, that date would also be my new niece's birthday? I gauged that the odds of this all happening could be millions to one. In fact, it was 76 million, six hundred fifty thousand to one.

Calculators do not lie.

But all I knew at the time was that it was way beyond coincidence. I told a few friends about it and they agreed that it was unexplainable.

"Your phone number matches your hire date?" they said skeptically. "And...you didn't somehow ask for that number?"

It was unexplainable – like some of the things I first encountered living in the Quad Cities:

"If this place is called the "Quad Cities," why are there *twelve* cities?"

"Why does the Mississippi River flow east-and-west in the Quad Cities instead of north-and-south like I had learned in school?"

"Why are there five big bridges across the river but the locals avoid crossing at any cost?"

The answer to the third question I quickly learned. You see, what I viewed as a river, the locals saw as a rivalry. The city straddled two states. The Quad Cities was all about Iowa vs Illinois. No matter the sport, the big game was always "The Hawkeyes vs the Illini." The Iowans bragged that their drinking age was only 18 while in Illinois it was 21. The Illinoisans countered that the letters in "Iowa" should stand for "Idiots Out Walking Around."

It was all in fun, of course, and you could argue that my life sounded like it was fun. And it did start out that way. But it was not long before I began getting signals – subtle at first – that my career was on the launching pad but it was not launching. It wasn't that anything was going wrong, exactly, it was just that there were career moves happening for my peers that were not happening for me.

My goal was to become a program director – a radio station's product manager - arranging the music, personalities and format. I could not think of a product more dynamic to market than a radio station. The craft is learned apprentice-style and generally a program director rises through the ranks, starting as an announcer. I began learning the trade at a student station in Indiana and then was hired by a "real" station in my college town to do weekend shifts. That gained me enough commercial experience to apply for a fulltime announcing job when I graduated – which ended up being the 7 pm-midnight slot at the little station in Moline. After a few years the staff got shuffled a few times and I ended up in the program director's chair.

I quickly found out that the radio industry is anything but static. Our afternoon announcer left to become the program manager at the popular competitor across the river. He took with him our affable, part-time, overnight announcer to do his morning show. Against all the odds, that unknown overnight guy would rise to become the most ascendant Midwestern personality of the decade. Listeners and advertisers flocked to our competitor's morning show and their ratings shot through the stratosphere. As a cub program director, all I had to do was to find a morning talent – not equal to – but substantially better than the best there currently was. Unfortunately, my station had less wattage, less prestige and far less money to offer.

I did give the effort a solid three and a half years of 12-hour days before I finally blew out – at 25. I told my girlfriend that I did not know what was going to happen next: I might have to change jobs; I might have to move; I might have to go back to school. She saw that she was not part of my long-term picture, and she was right. No one was.

Sadly, all of my education capital had been poured into the broadcasting bucket and I was discovering that the bucket was full of holes, despite every ounce of my interest, time, youth and energy. Why? In school, if I had given some class my full effort, usually I got an A or a B. But in my career so far, my work was barely passing.

This was the real back story when that "mountain-top" moment went down at Big Sable Point. To my friends I might have looked like someone who was going somewhere. But as I lay atop that sunny dune, I was just realizing that I was actually going nowhere. Yes, I was miserable. What had I done to screw everything up? How could broadcasting have been such a bad vocation if I had such a keen interest? After a few days my friends started home, but before I left, I went for a long stroll on the beach. It was then that I recalled my stumble through the darkness. Had I ever walked out of it? I had arrived at my career destination - just as I had arrived at my campsite that foggy night – and what awaited was an unhappy surprise.

I was in the middle of these ponderings when I saw it up ahead. I recognized its graceful form arching over the sandy parking lot. The sports car was long gone but the birch tree remained. I stopped beneath its boughs and looked up, recalling the afternoon when, from my dune perch, its contrast with the brightly painted car beckoned. Now, instead of the sports car in the parking lot below, it was me.

It was then I knew. I had chosen the career path that was supposed to make me a flashy item. Except, it hadn't and I wasn't.

The sunset over the water was peaceful, but I was not. Like a car stuck in the sand, I was rocking and spinning my wheels. I was only digging myself in deeper. If you have made a wrong turn that cost you dearly, this writer understands. A shiny object seemed a worthy prize at the time. But I was discovering that the road to find myself, was instead, leading me to waste myself. Meanwhile, I was not making any progress solving the bigger mysteries.

All the while, the little tree patiently awaited contemplation.

As I drove back to Iowa, I'll admit I didn't really know what to do next. Perhaps I needed to start over? That wasn't much of a plan. I found myself tapping softly on the General Manager's oversized office door as it slowly swung open. My boss was on a phone call but he waved me in. It seemed a long way across his carpet to the chair that faced his desk.

He hung up and turned toward me, removing his reading glasses. "Yes," he said with a smile and raised eyebrows. He was a gregarious sort who had come up the ladder through sales and advertising.

I started right in, "I'm really sorry to tell you this but I don't think I can be the program director here any longer..." I watched his face as I spoke. At first, he looked suspicious, like my physics teacher. Then, his expression told me he was not overly surprised.

Without breaking stride, he inquired, "So, what are you planning to do?"

I divulged that I didn't have another job or some grand plan. I told him the truth; he returned the favor by telling me he needed my help keeping the rest of the staff onboard through coming transitions. Most importantly, he told me that I could continue to be employed there. Though I had just taken a massive step sideways, I also felt as if I had just moved forward a bit too. I sensed that I had made the right tactical move, but I had absolutely no long-term strategy.

If you are in a time like this, just stop if you are only getting in deeper. It's not a time for quick decisions. It's time for right decisions. So, when I had a few days of vacation, I took an October trip up the Michigan shoreline to a spot on the map I had been wanting to visit. It was a place where I absolutely knew I could be alone to think.

Look closely at Michigan's Lower Peninsula "mitten" and you will notice that the northern-most finger-tip bends sharply to the west, like a tiny pennant in a stiff breeze. That's Waugoshance Point. At the very western tip of that pennant is the Waugoshance Cabin. There, I thought, I could determine a new heading. That one-room log structure has no electricity, no running water and no indoor plumbing. Its amenities are an outhouse, an axe, a pile of firewood and a hand water pump.

The cabin turned out to be a great place to think, and I discovered that thinking is really about all one *could* do there. 100 feet to the south is the shore of Lake Michigan's Sturgeon Bay; a hundred to the north and I was looking across the Straits to the Upper Peninsula. I couldn't go west because the Point is, well, pointed. It was all water in that direction. So, the only place I could really go was east, the way I walked in. I could go back down the mile-long path to the trailhead where I left my car.

As it turned out, none of the thinking I did at the cabin produced any grand solutions. What did, however, was the Point itself. I realized that I was at a point in my life (literally) where I could go no farther. I sensed that I needed to retrace my steps; to go back and pick up things I had lost. Whatever they were, they must have been important because it seemed that without them, I had reached the end of what I could do, both in my career and personally. I needed to turn my life around, but to do that, I first had to turn myself around. How exactly does one do that?

I started back home.

The return to Iowa at the 55-mph speeds of the 1980's made today's 8-hour drive take 11. I had thrown a camping tent into the car, thinking a one-night stop-over at Big Sable Point would break up the drive. It was around dusk when I arrived and all the roadside stands that had sold campfire wood in the summer had folded. I was forced to scrounge by flashlight for sticks on the ground.

As I cooked my hot dog over my micro-blaze that evening, for the first time in a long time I chanced to let my mind consider a "what if." I recalled that my student radio station had been powered by miniature transmitters that were located in the various dorms. They sent the radio signal into the buildings' electrical wires. "So, what if a state park had such a transmitter connected to its underground electrical system to create a 'camper radio service?' It could inform people like me about helpful things like...like where to buy firewood in the off-season."

Indeed, the retracing had begun.

At daybreak I had time to squeeze in a short hike before the day's drive. In sneakers, jeans and a rumpled sweatshirt I set out along the Ridge Trail, a sandy, needle-strewn path that rises and falls along a primeval shoreline. Before the campers, the loggers, the fur traders and the earliest inhabitants arrived, they say the waters of a larger lake had lapped at the west edge of that ridge. It would become the defining feature in the state park that would be built on the Point long after the lake receded.

From a high point on the trail, I gazed out across the tops of the autumn-tinged oaks and maples. A gusty breeze swayed some of the limbs as others looked on stoically. It was on that hike that I encountered another birch tree, and it occurred to me that this tree and the one in the beach parking lot might very well be offspring of a common ancestor. Or perhaps, one tree had produced the very seed from which the other had sprung.

Then I considered how uncountable is the number of individual lifeforms which exist on the earth at any given moment; and how impossible it would be to know the number of living things that have existed on the Earth throughout all of time. The number would be staggeringly large. And every one of them – I had been taught – from the largest elephant to the tiniest bacterium – had a "parent" plant or animal from which it obtained its life. That is, every one except one the very first one.

That thought stopped me in my tracks. The very first instance of life on Earth had ...no predecessor?

I was well aware that scientists have no explanation for life's beginnings. And yet the ones that wrote my biology text books had stated emphatically that the very first life on Earth – presumably some form of single-celled life – did *not* have an ancestor. It had no parent. They were certain of that. So, retracing my steps, I was asking a very basic question that should have occurred to me way back in high school: given that every life form that we have ever encountered originated directly from other life, if researchers admit they don't know how the first one originated, how, then, can they be adamant that it sprung from a non-living "source?" Given what we know of life, wouldn't it be not only possible – but probable – that Earth's first living thing had a *living* source?

And why were scientists stating with such certainty that it didn't?

It was on that particular October morning – standing on the trail on that ridge – that I began to seriously question this "reality" that had been presented to me. Who was tilting the table to favor inanimate beginnings? Who would stand to gain if I assumed that life had originated from an avalanche of accidents instead of a living ancestor?

These were only fleeting questions, but apparently, increasingly awkward ones. And I was about to find out they were not going unnoticed.

The Wide-Open Spaces

When I arrived home the weather had snapped cold. A frosty fog had draped itself across the river valley. It had been 3 1/2 years since I had crossed that river and I was well aware that I was spiraling like a pilot who had lost his orientation. I sensed that time was limited and there would be no warning when it was up.

After resigning my management position, I was assigned to work for one of my hires. Thankfully we shared respect; that made the transition less awkward. The company had also taken on two new salespeople. When I returned, their welcome notices were tacked to the bulletin board in the coffee room. I never did run into either of them.

But as the holidays approached, a new note appeared that announced a party. Everyone was invited. It was to be the Friday after Thanksgiving at the home of the new sales lady. I decided that I needed to go, but how would I explain myself? "I used to be the company's program director...until I resigned. But I still do work there. My plans? Well, I am working on those right now, actually..."

Thanksgiving Day was quiet and dark. New snow dusted the ground. A spot of Friday morning sun turned it to an icy crust. I found myself looking forward to the party, but I was absolutely unprepared for what was about to go down. Yes, nature abhors a vacuum, but this time it would be filled by someone way more unexpected than Father Garrigan.

I arrived as the day was fading. The house was a two story, square affair with old clapboard siding. After street-parking, I crunched my sneakers across the frozen yard to the front porch. A sudden breeze rattled the leaves that clung to the oak in the neighbor's yard. Then it was still. I rang the bell and as soon as the door opened, I realized that I may have been the only guest who didn't get the memo about going to the back door.

Inside, I didn't know a soul. In the living room was a spirited conversation among guys, a couple of whom looked a bit annoyed at my arrival, but not too annoyed to ignore me. Reasoning that the hostess would be in the kitchen, I made my way through the dining room where I dropped off my food item on a table scattered with snack bags and beverage puddles.

I didn't know anyone in the kitchen either, but upon seeing me walk in, the hostess broke away. She was not sure who I was and I had not met her, so we exchanged names. I thought she was about to take me around to her friends, but instead she came close and whispered, "I hope you don't mind, but there is someone here who has been waiting to meet you."

I was puzzled, and I will admit, a bit flattered. That kind of thing never happened to me. She walked me back into the dining room to a couch that was backed up against the wall. There, alone, sat a dark-eyed girl who looked way too self-assured to be lonely. "This is my friend Morana," she continued.

"And you must be Bill," offered the woman with the Mona Lisa smile. The hostess glanced Morana's way and started back to her kitchen guests. As I sat down, I was thinking, "How odd to arrive at a party where I know absolutely no one, only to be introduced to a stranger who appears to be expecting me." As I made pleasantries, I wondered how this girl knew my name when the hostess didn't even know it.

That should have been my first clue.

"Maybe my friend told you," Morana began, "you and I really need to talk. I've got a very important matter to discuss with you."

"And what matter would that be?" I parried.

As she began her response, I was distracted by her chocolaty eyes that laughed, almost out loud, as she spoke. They were distracting in a hypnotic sort of way. Her voice was calm and crisp; her speaking pace remarkably even. Coming out of media, it caught my attention. It was almost as if she was reading, her words were so precisely pronounced. She was clearly a pro, but a pro at what?

With embarrassingly little effort, Morana convinced me that the party was not the time or the place to discuss important matters. Instead, she proposed we would get a bite to eat after work the following week. The day was agreed upon, the 6th of December, at 6 p.m. at the Governor's Tavern. We exchanged phone numbers and when she looked at mine, she glanced at me as if she was going to say something. Just then, Morana's fiancé walked up and introduced himself. It was a bit awkward for me, but he appeared to be quite at ease, giving no impression that he was suspicious about her making meeting dates with guys she met at parties. They both acted like it was any old Friday night. I was beginning to suspect that it was all a well-rehearsed invitation to an investment pitch.

The Governor's Tavern was the epitome of discrete dining: low lighting with soft, dark carpet. Amber path lights directed one's steps. The tiered dining area was illuminated by the soft glow of green banker's lamps placed at each table. The man behind the bar sported a tailored vest with a tie. The Governors was hushed and polite and there was always a tinkle of piano wafting about. It was absolutely the right place to discuss important matters.

Morana was already seated at a table for two when I walked up. She looked at me as I approached and smiled kindly. After ordering an appetizer and making some small talk, she got right to her main topic. Her main topic was me. I immediately got the message: I was in a counselling session and it would be Morana running it. She paused for a moment and looked aside. Suddenly, her expression changed to amusement. "You think about funny things almost all the time, don't you?" she observed, her eyes sparkling. That was her opening, but it gave me no clue where she was going.

"Well, I am in the entertainment business of sorts," I responded.

"True," she said carefully. "But the business is not very entertaining for you anymore, is it?"

And how the heck would she know that? I wondered if she was a recruiter for some other radio company. Or was this the introduction to the "investment opportunity?"

Before I could respond, she looked directly at me and continued, "It's not really the radio business that's got the problem, though, is it?" I had been expecting some sort of acknowledgement of my public persona but not a character critique. "You are all about you," she declared, her eyes pointing at me like pair of darts, "and you can't see what's going on right in front of you."

I wondered if she hazed everyone like this to soften them up for the pitch. I was about to dismiss her and try to move the conversation to a more comfortable topic when she interjected, "Your father's passing when you were a boy was a defining event for you, wasn't it?"

"Yes, that really was quite a shock," I responded, playing for time. Her comment had put me back on my heels. "So, how do you know about any of *that*?" I stumbled. My mind was racing. Whom could she have talked to in the Quad Cities that would have told her about my dad? He had been gone for more than 10 years. The short answer was – nobody. Only my family in Ohio could have told someone about those events.

But Morana didn't answer. She just looked up from the table. "My dad died too when I was little," she offered, "and like you, there were some very odd circumstances associated with it."

No, this was not some sales presentation. Morana had inside information about me from someone, somewhere.

She began to look into space as she continued. "You are fascinated with the night skies, the deserts and the wide-open spaces," she stated. "But danger lurks in wide-open spaces. You really must be more careful. You are wandering. You are looking for a place to call your own, but you are trying to find it by imitating. Don't you understand that you are unique? Everyone is. But you don't understand that and you are floundering. Yes, you are failing."

I was way beyond dumbfounded. Where was she getting this stuff? All I could make myself do was to listen to her describe things that she could not possibly have known about.

Then she began to get specific, "Even your telephone number has a meaning. Don't you see that?"

"Well..." I started.

"And how about the dark-haired girl from Iowa that you dated...." She paused and looked at me as if I was going to say something. "That girl is hanging her head so sadly," she said. Morana was staring through me as if she could see her standing behind me. "You would not marry that one because there were no lights on her street. And the short girl with the blonde hair that you left when you moved here from Indiana. It's even better you didn't continue with her," Morana said, looking up at me cautiously. "She would have left you when you lost your job."

"Lost my job?" I repeated.

"Not yet, you haven't," Morana inserted. "But you will land on your feet. I see you in a room with rows and rows of circuit boards."

All I could imagine was that she had some source of inside information about me, but why would someone set me up like this? And it didn't explain how she knew all the personal details about my past. Then, I noticed that just before Morana spoke each time, she would pause, and when she paused, she seemed to be listening to someone. "So, how can you know all these things about me?" I challenged. "Can you read my mind?"

"No, not at all," Morana stated in a matter-of-fact tone. "I don't need to be a mind-reader," she said, looking me straight in the eye. "The ones who followed you here from Ohio tell me all I need to know."

Time ground to a halt. What "ones from Ohio?" Morana had no way to know I was from Ohio or that I had recently lived in Indiana.

So, there I sat, a person who got paid to talk for a living, yet as I looked across the table at this stranger who had just performed the impossible, I was unable to muster any meaningful response. I was so out-matched, I decided to put it back in her court. "I've obviously gotten my life turned around backwards, so where do you think I should go from here?" I queried. As the words left my lips, it occurred to me that I was actually asking a perfect stranger for personal advice.

"Well," Morana began, pausing to look away and then back with an intensity. "You are going to have to continue your search because until you find what you are looking for, you will just wander."

"And how will I do that?" I probed.

She was quiet for a long time. Then, as if someone was going to overhear her, in a low voice she said, "You are going to discover a very old book in the library that has the answers you seek. It's a book you have never read. But you are going to read it cover to cover. Only then will you gain the power to change."

"This old book is here in our local library...like the Davenport Public Library?" I questioned.

But Morana was not answering. Her face became distressed and her eyes teared up. Something had gone wrong. She stood up and pushed back her chair. Pulling on her coat, she wiped her eyes, gave me a half-smile and just said "I really have to go."

Without another word, she started for the door. I made after her as she made a beeline across the parking lot. "Morana?" I called. "Are you okay?"

She looked back as she walked and repeated, "I really have to go," now in a more frightened tone. She did not explain. She only proposed, "Maybe we can talk again sometime soon." But she said it as if it would be okay if "sometime soon" never came. She got into her car and drove off. I stood flatfooted in the Governors parking lot wondering what I had just experienced. I went back over the conversation again and again and couldn't think of anything I had said that should have upset her. But it would not be long before I would be the one upset.

I drove back across river to my apartment. Nothing like this had ever happened to me, or to anyone I knew for that matter. That meant there was going to be no one I could talk to about it. If someone had called me up to tell me a story like this one, I would have dismissed them as crazy – like someone who sees ghosts or UFO's. But I knew first-hand that it wasn't crazy... and I wasn't either.

When I arrived home, I sensed that something was amiss. I sat down to process it all and it was not long before I knew what it was: I apparently hadn't been alone for the last 3 ½ years – or ever – like I thought I had. Who were these watchers who whispered in Morana's ear? This was not trickery. I had seen her waiting on their every word. And they came *with me from Ohio* no less? It was so bizarre that it just couldn't be real, and yet Morana had proved her case beyond any shadow of a doubt. Then it occurred to me that there existed no one person on the entire planet who was aware of all of the personal details that she had offered about me – not even someone in my family or my very best friend.

So, just who had been talking to her?

The whole incident left me numb. It was as if someone "not from here" who had known me from my earliest days was feeding her my most personal information. It was beyond frightening. I barely slept that first night. And the more I considered it, the more I realized how vulnerable I was. Whomever "they" were, they had every advantage: they had been totally undetectable while I was totally exposed; they were knowledgeable about my past – and supposedly even about my future – while I had no idea that they even existed until that night.

Was I being observed by someone invisible who had superior powers? Such a thought could make one paranoid. But, by definition, paranoia involves delusions, and what this Morana pulled off proved to me that these watchers were not hypothetical. They were very real.

If I had told any rational person about this, I would absolutely not have been believed. Would you trust such a story, even if it came from your very best friend? So, I sort of withdrew and went through the motions. I kept thinking how drastically different reality must be than I had originally believed. Are these "ones" some sort of ghosts, evil spirits, angels, aliens or what? And why is no one aware of them except people like this Morana?

I made an attempt to contact her a few days later but she no longer wanted to be reachable. Something had obviously gone sideways. I stopped by the bookstore where she told me she worked. She was polite but noncommittal. She was clearly reticent to reengage and reticent to explain what had gone so wrong.

Interestingly, of all the things we discussed that December night, it was her observation about my interest in the night sky that I could not shake. No one – and I mean *no one* in Iowa or Illinois – and certainly not some stranger – knew that I had a longstanding interest in astronomy. That - and her description of me as self-centered – with the cryptic warning about wandering into the "wide-open spaces" – really struck a chord. Had I been seeking my fortune far from home for a reason? Perhaps I wanted to be free from the constraints of family, friends and faith so I could do as I pleased? In doing so, had I wandered into a precarious place?

The old fluorescent fixtures hung on rusty chains from the ceiling of the Davenport Public Library. They cast a stark light across the stacks. I identified a set of shelves with hundreds of titles on all manner of topics: philosophy, religion, self-proclaimed prophecies, spirituality, clairvoyance and even aliens and the paranormal. It was there I began my search for the "old book" Morana had recommended. But why was she so vague in describing it? She had been so uncannily accurate with her observations about everything else. Yet, I had no doubt that despite the daunting number of dusty titles before me, I would know when I came across the one. Interestingly, with each book I looked through, I became more convinced that the one I sought was not about someone's subjective experiences or their personal powers.

There are shelves of books by writers claiming to have contact with unseen personalities, reminiscent of Morana's "ones from Ohio." The extensive writings about Edgar Cayce's "readings," which began in the 1930's, are a prime example. Cayce – known as the "sleeping prophet" – would go into a trance, hear voices and provide detailed cures and healing remedies to people who would take dictation. When he awoke, he would not be aware of anything he had said.

After flipping through many of the titles, I became convinced that writings like these were so subjective that anyone could pick up a pen and paper and start his own philosophy or cult. How tempting it would be for a person with impressive other-world connections – or just a charismatic personality – to gather a flock of faithful who would come to rely on them for direction. Had I never met Morana, I would have dismissed all books like this as outright frauds. But you can see why I had to allow that some of them probably do contain sincere depictions of actual contacts people have had with something or someone unseen. But although it's obviously possible this could happen, how would a reader know if the authors' claims were real? Were there really "ones from Ohio," or was that a fabricated reality? And how could one be sure that the authors' spiritual "connections" were well-intended?

One could get lost in such territory.

Then, out of nowhere, a message appeared on my answering machine from Morana, asking if I would contact her. When I returned the call, she asked if I would like to attend a meeting she thought I would find "enlightening." She didn't provide details. She just gave me an address of a house in Moline and a time to arrive.

This would turn out to be another eye-opening evening. Morana was already there with a half dozen others, seated on metal folding chairs circled in the living room. They began with a discussion in which people took turns giving updates on their favorite spiritual topics. Most talked about books they were reading. Others believed they had special abilities, like one older gentleman who said he could tell someone's future by "reading" the tea leaves in a cup. Another lady read palms.

Creepy? Oh yeah. I was full-on squirmy.

Morana proceeded to introduce me as new friend who was interested in learning. [To use Harry Potter jargon, I was a "Muggle" – someone who had no obvious magical abilities.] That seemed to get peoples' curiosities aroused. I listened awkwardly for a while, but eventually they started asking me questions. It was not long before they discovered I didn't know the first thing about their topics and the closest thing I had to a "spiritual connection" was that I had been raised a Catholic. Surprisingly, a number of them were interested in talking about that. It seemed everyone in the room had known a Catholic at one time or another. The palm-reader lady said that she had been raised Catholic herself until she became "enlightened." Great.

I shared with them my interest in astronomy and commented that the Catholic Church and science didn't have a very good record of being on the same page down through history. A quizzical look came across Morana's face. She interjected, "But you know, the very earliest scriptures actually align very well with science." People nodded agreement. "Have you done any research on that yet, Bill?" I sort of knew what she was getting at. I told her that I been down at the library a lot. She cocked her head, smiled kindly and gave me her "sad puppy" look.

When the meeting was over, people adjourned to the basement to catch the latest installment of "*The Thornbirds,*" a popular TV miniseries about a wayward priest and a female parishioner. It was really too much. I excused myself and drove home.

I will say that the people in Morana's group were sincere about what they believed, but I came away with the impression that most of the attendees – including Morana – viewed themselves as brokers within their own spheres of influence. They went out of their way to describe how they had given counsel to friends and family and how much their subjects had benefitted.

As it turned out, that was the only meeting like that I attended. I wasn't interested in developing any personal powers, and frankly, I didn't see a Ouija board in my future.

Meeting Morana had made me wary of wandering into the wide-open spaces.

The Clue in the Cluster

The chains were still rusty at the library. Despite my misgivings about Morana, I never doubted that she was trying to coax me in the right direction. It took a few more visits, but I was eventually able to locate the book which I thought she had been directing me to. Her comment about the "scriptures" proved to be the key. What was puzzling was why she had avoided just telling me that the book was – in fact – *a Bible.*

Perhaps she knew I would resist opening up a *Bible*? She was right. I didn't own one, I had little interest in it and I had been avoiding anything related to Western religions because I associated them with the Catholic brand. But many of the books I was finding referenced those early writings. It made sense because – as I learned – the ancient Hebrew scriptures contain the oldest source documents with explanations that claim to describe man's beginnings. Some fragments date back thousands of years and many modern religions use those original writings as their foundation and then springboard from there.

But I wasn't interested in deviations or denominations. I wanted to know what the earliest texts actually said so I could judge for myself. Here is where it began to get interesting.

One of the earliest passages I found was in a book named "Job." What first caught my attention was that the book, written in something like 500-1000 BC – contains science facts that are totally out of synch with the knowledge base of that era.

The main character named "Job" is presented as a historical person who undergoes a series of severe trials. The testing is permitted by a character identified as "God," after a challenge by his adversary "Satan," who asserts that he can cause Job to curse God's name if Job's circumstances are made sufficiently unpleasant. God says it's not going to happen, and to prove it, allows the test to proceed. So, the contest is on, and Satan is permitted to temporarily deprive Job of his wealth, his health and even his family, leaving him alone to process a truly awful situation.

To make matters worse, Job's friends show up one by one to give him unsolicited advice. Perhaps his associates believe that because they have not suffered such catastrophic losses, they have the right to tell Job what he should have done differently to avoid his circumstances. Through all his trials, Job steadfastly refuses to place the blame on God, to Satan's chagrin. As a reward, God restores Job beyond what he had to begin with. Finally, in the stunning conclusion, the God character shows up, speaking from within a windstorm, and audibly calls out Job's "counsellors," asking pointedly, "Who *is* this who darkens counsel by words without knowledge? Now prepare yourself like a man; I will question you, *and you shall answer me.*"

He continues, "Where were you when I laid the foundations of the earth? Tell me, if you have understanding. Who determined its measurements? Surely you know!

Or who stretched the line upon it? To what were its foundations fastened? Can you bind the cluster of the Pleiades, or loose the belt of Orion?" (1)

The character identified as God directly challenges the "wisdom" of Job's friends. He characterizes it not as wisdom at all. He says they have spoken "words without knowledge" and calls their "wisdom" darkness. Then he proceeds to challenge them with a series of questions only the creator of a universe could know the answers to. Instead of using scientific terms, he uses those of an earthly builder: measurements, lines, foundations. He speaks of the Earth as if it was a house. He effectively says to Job's counsellors, "based on your limited knowledge, how could you even attempt to build a home like this Earth?" Then, in a demonstration of superior knowledge, he calls out the stars – specifically the Pleiades Star Cluster and the three belt stars in the constellation Orion.

The reference to the stars made me sit up and take notice.

It's critical to understand that when the Book of Job was written, people did not have any idea what stars really were. They knew stars to be bright dots of light in the night sky, but they did not know if they were pinholes in a black background or bright globes hung from dark ceiling. [Those actually were two popular theories of their day.] But ancient people had no way to know that stars are massive, burning balls of gas, scattered in a vacuum at vast distances, influencing nearby bodies with their gigantic gravitational attraction. People would not have an inkling of what stars really consisted of for thousands of years.

But what shocked me was the writer of the book of Job clearly knew what stars are, and specifically, that they have gravitational attraction.

If you go outside on a clear winter night and look up at Orion's belt, you will see three bright stars in a straight-line grouping. Not far away in the same sky you can spot the Pleiades Star Cluster in which, with good eyes, you can make out about seven stars in an irregular pattern. Both Orion and the Pleiades are obvious sky features and were well known to the ancients.

But read carefully: The character called God refers to the Pleiades as a "cluster," while the belt stars in Orion are not identified as a cluster, despite both features visually containing roughly the same number of stars. Why would the writer have referred to one as a cluster but not the other? And I noticed that God asks, "Can you *bind* the cluster of the Pleiades or *loose* the belt of Orion?"

It would be 2500 years before astronomers would discover that the Pleiades consists of not seven, but more than a thousand stars arranged in a tight cluster formation. Even more recently they found that every one of those stars is travelling together as a cohesive group, each attracted to one another by mutual gravity. So, the Pleiades is literally a bundle of stars, bound up by its own gravity, moving through space as a unit – a cluster.

In stark contrast, the "belt" in the constellation Orion, which also consists of a small grouping of bright stars, has virtually nothing in common with the Pleiades. The three belt stars only *appear* to align and have proximity to one another because of the perspective we have from Earth. In other words, the three are not in a physical line at all; they only appear to be, due to our viewpoint. More importantly, the stars in Orion's belt are not remotely close to one another and certainly are not gravitationally attracted. In fact, the three are travelling in totally different directions in space.

These writings caused me to pause and take a big step back. So, the character identified as "God" in these 2500-year-old passages, clearly cites as evidence that he is the universe's creator, his personal knowledge that the Pleiades stars are bound together while the stars in Orion's belt – which the ancients had always assumed to be bound like a belt – are not bound at all – but totally loose. Obviously, no human at the time Job was written had a way to know that stars have gravitational attraction. The fact that heavenly bodies could be attracted to one another by gravity would not be discovered for millennia.

I knew that these writings were not intended to be a science book any more than science books are intended to impart spiritual truths. And yet, it seemed that I had come across recently unearthed scientific facts in a book that predates the word "science" itself.

Even more stunning: The character identified as God states that he is not only aware of this, but he is *personally responsible* for the arrangement, claiming that he himself did the binding of the Pleiades stars and the loosening of Orion's. It's one thing to make the claim to be the creator of the universe but it's quite another to make statements that, on the surface, would appear to prove it.

And this was not the only instance like this that I found among these ancient writings.

The Parallel Progressions. In the first chapter of the Old Testament – the book of Genesis - everyone in Western Civilization knows that there is an explanation for Mans' origins in which the creator launches the universe with the command "Let there be light." I had always assumed that this was the claim to the creation of the sun and sunlight. But on reading carefully, I saw I was incorrect. It actually says the universe existed well *before* there was light.

The first words in Genesis clearly state that before our sun existed, there was already light in the universe, and even before there was light, there was a long period of total darkness. It reads, "In the beginning, God created the heavens and the Earth. Now the earth was formless and void, and darkness was over the surface of the deep.... Then God said, let there be light, and there was light." (2) In other words, Genesis says that right after the universe came into being, there was a period of total darkness, and during that time, the earth had no form. Only after that, came light. Then later in the timeline, actual stars and planets appeared.

That is exactly the opposite of what I thought it would say.

But then I recalled that this progression is precisely what today's astrophysicists teach - that for the first 300,000 years after what appears to be the "beginning of the universe," there was no light at all and no elements as we know them. [Recall that the only thing they believe existed was a formless kind of matter.]

Today's best science confirms that only after this lengthy period of darkness did the first radiation in the form of photons (light) appear. The release of the first light energy was so intense that it can still be observed today, coming at us from space from all directions – a mysterious type of energy scientists named the "Cosmic Background Radiation (CBR)." Only well after the release of this first light would conventional light sources such as stars and our sun have formed.

Amazingly, both scientific observations and these ancient writings align precisely on the very same sequence: 1) darkness with formless material, 2) the first light and then 3) the formation of conventional light sources such as the stars and the sun.

Then the Genesis account proceeds to describe:

_ The separation of energy from solid matter – allowing for the differentiation of light and darkness.

_ The beginnings of our planet and its seas and atmosphere; then dry land; the sun and the moon.

_ The first life in the form of plants, then sea creatures, birds, land animals, and ultimately, humans.

It had never struck me before how this very same progression – beginning with utter formlessness at one extreme and intelligent life at the other – is attested to by both the Old Testament and modern science.

To get some context, I researched the other earliest creation stories – those of Babylon and Egypt – and found that their explanations for our existence are at best fanciful tales, reminiscent of those from the Greek and Roman civilizations I recalled from school. The Babylonians imagined the first gods came about when fresh water and salt water were mixed. Those gods then gave birth to the multiplicity of other gods that would create and rule over nature. The early Egyptians imagined the first god created *himself* and then proceeded to mate with his own shadow to create all the other gods, each of whom were responsible for a different aspect of the world. But in stark contrast, the Genesis account describes a logical progression from formlessness to intelligence, paralleling what we have learned from direct scientific observation.

Image and Likeness These early scriptures go on to state that humans are the most advanced creatures on Earth and are unique among all the life forms because they were made in the "image" of the creator. This assertion squares with the reality that everyone acknowledges: though people share certain physical similarities with animals, it is obvious that humans possess abilities vastly superior to even the most advanced animals in terms of speech, reasoning, creativity, imagination, writing, self-determination, intelligence, learning and archival knowledge. Humans have an innate understanding of right and wrong not evidenced in animals. And most importantly, humans have the ability for highly developed relationships based not merely on feelings, but on love.

Genealogy. Then I came across the section of Genesis in which the writer asserts that the first woman on the earth, "Eve," would become "The mother of all the living." (3) It says all humans born on Earth would be able to call Eve their direct ancestor. I remember hearing that when I was growing up but I considered it to be just a story on the level of those the Babylonians and Egyptians told. But I shouldn't have been too shocked. It's obvious that all humans are related if you go back far enough. But to a single man and woman? Yes, it's obvious. Even Darwin's theories assumed that. Then in 1987, studies on cellular mitochondrial DNA (mtDNA) proved conclusively that all women on Earth are direct descendants of a single, common female ancestor who would have lived well before recorded history. Because all men are born from women, it is inescapable that a specific woman in our distant past was the common ancestor of us all. Science and the ancient scriptures concur that this woman was literally "the mother of all the living."

Vastness. After all the phases of creation were finished, the Book of Genesis states that the "heavens and earth were completed in their *vast* array." (4) Really? Once again, the problem is obvious: at the time of Genesis' writing, people knew nothing of the universe's virtually infinite size and varied nature. Yet these scriptures, written more than 2500 years ago, state that the heavens and the earth comprise a "vast array." Recall that it has only been within the last few years that people with telescopes began to understand that the distance to the stars is immense and the number of stars is uncountable. Even as recently at 500 years ago, they did not know that the Earth is a sphere, fully 8000 miles in diameter. Nor did people know until the last few hundred years that there are thousands of distinct varieties of animals and plants scattered across seven expansive continents. Did I fail to mention the invisible universe of unending detail revealed under the magnification of the microscope? Indeed, with all our technical abilities, we still cannot even begin to survey the vastness of space and the array of life that inhabits our own planet. Apparently, however, the author of these ancient writings knew about this.

After I encountered these passages – and there are many others not referenced here – it became overwhelmingly obvious that the one responsible for these ancient writings had inside knowledge on the very design of the universe. Because the writer also claims to be conveying the words of the *creator* of the universe, one suspects that these clues were woven into these ancient narratives quite intentionally to prove their veracity to modern readers.

In other words, only the one responsible for the universe could have inspired these passages.

But a world that was *created*? Really? Isn't that what all the religious zealots have been pushing us to accept? My mind instantly resisted. A created world would be a narrow world with moral consequences, right? Isn't a created world the very world we would want freedom *from*? Does anyone really want a creator to obey? A culture like ours, bent on unfettered freedom, has to reject these notions.

But I had to consider: could it be that because I came from just such a culture, did I prematurely dismiss the possibility? Believing in a created universe would put me at odds with reality unless, in fact, it *is* reality.

I had to admit, though, if the world actually was created, it certainly would untangle some of the science conundrums:

__ How an expanding universe comes into being when physics clearly states it's impossible...

__ How life could have originated from lifeless materials...

__ Why space is so silent when it should be a busy conduit for celestial communication...

__ Why, after millions of years, only one life form we are aware of – humans – is advanced enough read and understand these very words.

I had begun my life with a good amount of faith in science. But to my surprise, I was discovering that there is a good amount of science in faith.

I stared up into the starry expanse above Big Sable Point one evening and realized that the "Wheel of Fortune" puzzle that had confounded me was resolving:

I had understood that the universe was virtually infinite.

I was beginning to see that the one responsible for it would have to be even more than that.

Stepping Out of the Spotlight

Can you imagine what went through Truman's mind when he escaped the set and saw what was going on backstage? Did he marvel at the cameramen, the set designers and the directors scurrying about with their clipboards? And how was he going to deal with the real-world freedom that seemingly would present no boundaries? Would it have been easier to have remained in his manufactured world for all its ease, comfort and simplicity?

But the biggest reality Truman may have grappled with: *he* had been "the show," but could no longer be. There was no going back. Walking off the set meant accepting that he was one of billions of inhabitants of Earth, each with an equal footing. None of us possesses the inherent right to command an audience. His TV "life," though he didn't know it at the time, had focused on him, quite literally. Leaving the show would mean a drastic loss of the control and the attention that naturally accompany stardom.

"Who is this who darkens counsel with words without knowledge?" It was most humbling when I realized that the author of Job – the one who claimed to be my Creator – was talking not only about Job's friends, but about me. Ironically, my darkness was a direct result of my insistence on standing in an imaginary spotlight.

"Darkness," in the passage from Job, was equated with "words without knowledge" – the belief in things that simply are not true. Indeed, I had been intertwined in just such a web of worthless words. I worked at a radio company. We made our living saying things we thought would be entertaining but had no basis in truth: there was news that wasn't new; commercials that persuaded with half-truths; music that spoke of "love" as if it was a product for purchase. It was all neatly woven into a perpetual program stream for consumption by listeners who were daily encouraged to listen more so they could buy more. The underlying untruth was that buying more would bring listeners "life-changing" fulfillment.

But it went even deeper. Morana was correct, observing that I was self-centered. But that didn't bother me much because selfishness is an acceptable trait in our culture. I knew I could get a pass on that. What I didn't understand was that my myopic perspective originated from a deeper untruth that was blackening my path. Somewhere along the way, I had begun denying that there even existed a creator – a god. In my heart, I had made any god, at best, irrelevant. Instead, I had chosen to believe the lie that I was the only show in town.

As a flame attracts the proverbial moth, radio, TV, films, the stage and media of all stripes are known to attract those who seek the spotlight. I was not the only person I knew in my profession who could not see past his own nose. Now, fast-forward to today and see how amplified the "obsession with self" can become for some who carry an imaginary stage in their very pockets.

More than 50 years ago, seers in the entertainment industry had already predicted where it was all headed. Philosopher Marshall McLuhan famously asserted, "All media exist to invest our lives with artificial perceptions and arbitrary values."[94] Pop culture commentator Andy Warhol predicted "In the future, everybody in the world will be famous for fifteen minutes." (95) These early media prophets accurately foresaw the personal and social deterioration that would result when an "imaginary spotlight" became accessible 24-7. The internet and its agents would forge a high-speed highway directly into the minds of many. The result would be devastating - a pandemic of self-preoccupation.

This explains why you can find, in most every city, programs offering therapy for adults and children dealing with the isolation, obsession and depression – the darkness – that results from an unnatural focus on that which is not true. It should be obvious that any flat screen merely presents a two-dimensional world, while real life is multi-dimensional. But many are nonetheless entranced. Some violently act out their disdain for themselves and for others. We witness the results in the news every day. Meanwhile, the complicit media stand back, hands-in-pockets, asking society to provide the solutions. I speak from firsthand experience.

But what I had failed to grasp was that believing lies brought me not freedom, but blindness; seeking the spotlight had brought me not light, but darkness. "A man who walks in darkness stumbles,"(5) it is stated in the scriptures that I found. Nothing could be a more fitting analogy for my first 25 years. Then I came across a corresponding passage that asserted, "The fear of the Lord is the beginning of knowledge. (6) That's actually the very first proverb in the Old Testament book of the same name. It means that trust and respect for the Creator of the universe is the very cornerstone one must have for there to be a foundation of truth to build one's life upon. Knowledge of the truth was what my words had been lacking. I was beginning to see that I needed some of that.

I was tired of stumbling.

Until then, I had been of the mindset that acknowledging God meant adhering to a rigid set of rules. I had thought that it would limit what I could do rather severely, and honestly, it wouldn't be very much fun. Would I have to wear special clothing or swear off getting married? Would I have to be weird or boring to be "spiritual?" My images of faithful people probably originated with movies and TV shows that portrayed them as outliers and oddballs.

But there was something about the fact that for there to be a Creator, he would necessarily be infinitely powerful. That sold me. I had mucked things up impressively by trying to handle life my own way. So, I began to think about the possibilities should I gain an ally who was *that* kind of powerful. Was it even possible after all of my missteps? Yet, I was finding in the scriptures a description of a father-like God who didn't want to limit my future possibilities so much as show me a future that was limitless.

Now, if you are like me and have never cracked open the Good Book, let me be the first to tell you that what is there may be different than what you have been told. I was surprised and frankly, I couldn't believe I was even doing it.

Right away, I was curious to see how many of the Catholic Church rules were in those original scriptures. I discovered that the basis for the core Catholic faith is definitely there. But what I did not find was a neat listing of the seven sacraments or other expressly Catholic formulations like sainthood or the worship of Mary. Is meat-eating prohibited on Friday? It's not there. Is it a sin to miss the mass on Sunday? I found nothing about that. So, I went looking to see what *is* there and what I found was intriguing.

For one thing, I had never-before connected that if there really is a creator, then the creation must have a specific design. For instance, I recalled from my college psychology classes that it had been assumed that any animal or human could learn anything by the presentation of a positive or a negative stimulus. Think of an electric dog collar that applies a mild shock if the animal decides to venture outside a boundary. Scientists had assumed that this was a universal means for all living things to figure out how to adapt and live in their environments. And they found, to a limited degree, that is true. But what shocked the researchers [pun intended] was that some animals could not learn from specific stimuli while other animals, in precisely the same situations, could. For example, chickens ignored colored lights to clue them to which feed was not nutritious. Certain other animals learned the lessons about food quality using light colors quite readily.

It is now widely understood that all living things have hard-wired design constraints. This should not be surprising. If we are not evolutionary products, but instead, handcrafted creations, isn't that exactly what would be expected? The Natural Selection theory – the one in which "evolutionary forces" push the development of animals and plants only to the degree that they will survive – is way too blunt of an instrument to be responsible for the stunning array of life variation that we see on our planet. It cannot account for all the diversity because that hypothesis only requires a living thing to stay alive and have offspring to be retained genetically. That's a very low bar. Natural Selection, though it may be possible on a limited scale, could never be the driver of the gratuitous panorama we see in nature. It gives no style points for variation or beauty.

Instead, what if the Creator actually *intended* us to observe our diverse world and conclude that it could only have been created – not by an avalanche of accidents – but by an artist?

This was totally new thinking for me.

So, are there designs that govern how all life works; how human relationships work; how your life and my life work? The scriptures I found repeatedly illustrate that when we try to work outside the design boundaries, bad outcomes are guaranteed. We all know that there are behaviors that never turn out well: lies, slander, violence, hate, murder in its many forms, greed, envy, theft, etc. Everyone knows the expression "what goes around comes around." But it's even more than that, because spiritually speaking, whatever we do to each other, we also do indirectly to the one who created us. What "goes around" comes around *his* way as well. (7).

And that would explain a lot about why relationships in my life had never taken root. Sure, I was free to do as I wanted, but at what cost? If pursuing freedom meant me wandering outside the way relationships are designed to be – committed, male & female, monogamous, lifelong "til death do we part," – was my inability to grow up a result of my insistence on coloring outside the lines?

Another thing I had heard for years was that there was this massive conflict between the scriptures and science. But, as I alluded to before, it seems to me that if there is one, it's really a conflict for the reader who is going looking for it. For example, someone who cannot or will not accept that some scripture passages are literal and some are intended to be figurative is going to have a hard time interpreting meaning. Be wary of those who might try to convince you that the scriptures are totally literal or totally figurative. It's an easy proof that both extremes are wrong-headed.

I learned that the ancient writings use various means to convey the truth, and one of the very first things that I had to wrap my head around was that God did not say that he merely *told* the truth. I suppose I would have had no trouble believing that. But he went infinitely beyond that when he made the astounding claim that he *is* the truth. He asserted, "I am the way, *the truth* and the life."(8) - a transcendent claim that could only be made by someone who was either insane ... or ...by the Creator of reality itself. Only the latter was worth my consideration.

But the combination of infinite integrity and infinite power describes a being who is clearly worthy of something I did not know how to do – worship. I had worshiped myself and I had seen where that had gotten me. What would it be like to be associated with a being of infinite power and truthfulness? I was not prepared to worship a man. Why would I? But these writings made it apparent that this Jesus described in the New Testament was effectively a man *and God – too.* He refers to himself as the "Son of Man" and the "Son of God" in various places. If that's true, it must have been obvious to others that he was not a mere man just by observing his actions and listening to him speak. He attracted a massive following, which is why the religious elites were jealous and sought to kill him. Part of the reason for his appeal, the scriptures say, was that he did not speak like they did. Instead, it was said, he spoke "as one who had authority." (84)

I found a number of illustrations of this, but one of the best occurred when the Jewish leaders – the very ones who were seeking to discredit Jesus – sent armed guards to bring him in for questioning.

When the guards returned empty-handed, they asked them angrily, "Why didn't you bring him in?" (9) The guards didn't have a ready answer to excuse themselves. Instead, they replied candidly, "We have never heard anyone speak like he does." (10) One can only conclude that these men believed Jesus was exactly who he claimed to be – God in the flesh. After hearing what he had to say they were obviously more concerned about offending him than their superiors, despite the severe discipline that would have awaited them.

I became convinced that no mere man could speak like he did when I found a teaching that Jesus delivered to thousands of common folks like you and me. I can tell you that his words shouted to me right where I was - in a dark bedroom in a tiny apartment at the corner of Charlotte and Carey Streets in Davenport, Iowa.

" ... Do not worry about your life, what you will eat or drink; or about your body, what you will wear. Is not life more than food, and the body more than clothes? Look at the birds of the air; they do not sow or reap or store away in barns, and yet your heavenly Father feeds them. Are you not much more valuable than they? Can any one of you, by worrying, add a single hour to your life?"

"And why do you worry about clothes? See how the flowers of the field grow? They do not labor or spin. Yet I tell you that not even Solomon in all his splendor was dressed like one of these. If that is how God clothes the grass of the field, which is here today and tomorrow is thrown into the fire, will he not much more clothe you – you of little faith?

So do not worry, saying, 'What shall we eat?' or 'What shall we drink?' or 'What shall we wear?' For the unbelievers run after all these things, and your heavenly Father knows that you need them. But seek first his kingdom and his righteousness, and all these things will be given to you as well." (11)

The moment I read this, in my mind's eye appeared the little birch tree. For the first time, it was clear to me why it beckoned. That little plant had no job, no money for food and didn't have clothes to keep it warm against the brutal lake winds that whipped it relentlessly. Yet, in its tenacity, the tree was doing precisely what it was designed to do: it was in-synch with the one who created it. Indeed, it was very, very well cared for.

In contrast, there was me. Sadly, I recognized myself as the one of "little faith."

I had no way to know that I was about to have my job ripped away from me in a matter of months. Trusting the words contained in these verses would prove pivotal. Was my solution as simple as: "Seek first his kingdom?" Just how would I do that? And how would I behave if I was no longer the star of my show?

I got up and walked into the kitchen. The windows were open and the fabric curtains undulated in the warm spring breeze. The light of a new day was streaming in. I hadn't done anything different, but something was changed. I looked around. There was an unnamed "something" within me that was not there before.

I didn't know what to do next, and yet, for the very first time in my life, I knew exactly what to do.

The Creator and the Created

The Jews were certainly emphatic about it. Even I recalled from my Catholic days that there was supposed to be only *one* God – not a multiplicity of them. Worshiping multiple gods was pagan, right? And it makes sense that if the world was created, it wasn't created by hundreds of different gods. But, given that, how was this one God also a trinity? I did not understand that and I was not even finding the word "trinity" in the scriptures anywhere.

In contrast, the other ancient cultures had different gods seemingly for every day of the year. The Apostle Paul famously called out the Romans for even having an "unknown god"[12] on their roster of gods that numbered upwards of two hundred. Yet, I didn't understand how there could be three distinct beings that made appearances and yet the scriptures also stated that there was only one God.

I learned in my research that I am not the only one that struggles with this one. Apparently, it is one of those things that the finite mind may not be able to grasp. People have tried to explain that an egg has a shell, a yolk and albumen (egg white) and the three are all part of the same egg. That explanation seemed too simplistic, so I had to reset to "I still don't understand." Perhaps we will never truly understand because we can never be more than individual souls?

Later, though, I came across an explanation that made more sense to me. The clue is in the very first verses in Genesis: the Creator refers to himself in the plural. He states that he will "...make man in *our* image, in *our* likeness." [13] He is telling us in the very first sentences of the very first book that there is more than one aspect to his nature. And it's actually more than "aspects." It's more like "persons." We are introduced to the Father and to the Holy Spirit in various places. We also see it explained that Jesus is the "Son." It is made clear that these three "individuals" are all God, all have been present eternally and all are responsible for our existence.

But it's even more than that. The scriptures also say that love is shared, first and foremost, among the Father, the Son and the Spirit – and God then shares his love with the rest of creation. That explained to me why, when I first heard that "God is Love," (14) it struck such a resounding chord. Love is what God always *is* and therefore love is what God always *does.*

Jesus explained:

"Very truly I tell you, the Son can do nothing himself; he can do only what he sees his Father doing, because whatever the Father does, the Son also does. For the Father loves the Son and shows him all he does. Yes, and he will show him even greater works than these, so that you will be amazed. For just as the Father raises the dead and gives them life, even so, the Son gives life to whom he is pleased to give it. Moreover, the Father judges no one, but has entrusted all judgment to the Son, that all may honor the Son just as they honor the Father. Whoever does not honor the Son does not honor the Father, who sent him." (16)

In another passage, the disciple John writes, "In the beginning was the Word (referring to the Son). The Word was with God and the Word was God. He was in the beginning with God." (15)

And Jesus advised about the Holy Spirit, "...When the Helper comes, whom I will send to you from the Father, the Spirit of truth, who proceeds from the Father, he will bear witness about me." (17)

These passages illustrate the graceful, respectful, affectionate nature of the persons of God. The Father and the Son and the Spirit are shown to be "one" in their love for each other and for us. What a different picture than I had been shown. Hardly an authoritarian, love is what God *is* and his love is *the* model for how we relate to him and to each other. This is love in its perfect state.

And it's consistent with his loving nature that this God – who has infinite power and is also both imminently trustworthy and loving – maintains control very differently from certain authorities on the Earth. The expression goes that "power corrupts and absolute power corrupts absolutely." A dictator on the earth who has absolute power has no reason to be truthful or loving because those characteristics often don't benefit him. In fact, they could cost him power and control. Instead of dealing in the truth, he spins the narrative; he doesn't spend a lot of time figuring out how to love his opponents, often finding it more expedient to imprison or kill them. But quite unlike a human dictator, God is omnipotent *without* resorting to lies, hate and manipulation. He does not have to. Unlike us, he holds all the cards and tells us that in no uncertain terms. In the book of Isaiah, he explains, "My thoughts are not your thoughts, neither are your ways my ways." (18)

Despite us being made in his very likeness, he is very different from you and me: his love is perfect and cannot be adulterated by the imperfect. (19

And speaking of imperfection, there's those Ten Commandments. Everyone has heard of those. I had always assumed that they were provided to perfect us, but I learned that quite the opposite is true. The Apostle Paul points out that the purpose of the commandments is really to illustrate that relying on our own resources, "perfect" is precisely what we cannot become. (20)

And Jesus clarified that the ten can be summed up in just two: to love God and to love others with all our hearts. All the other commandments are corollaries of these. (21)

Now, I had always calculated that I had some wiggle room when it came to commandments because I had never killed anyone or stolen their things or run away with their spouse. Sure, I might have lied a few times and been envious of some other guy's car, but that was about the extent of what I could think of that I had ever done wrong. So,

I assumed I would shake out relatively blameless compared with the world's murderers and thieves. But wouldn't it be convenient if I could assign myself to be my own judge, jury and defense attorney? Certainly, then I could proclaim myself to be innocent and celebrate my acquittal.

But by clarifying that perfect love is the standard of a perfect God, even the very kindest of us must admit that we cannot stand blameless. No matter how well we think we love other people, we always could have done better. That deficit goes on our account. When compared against God's gold standard of love, even the best of us has an impressive list of priors.

I discovered an amazing story that Jesus told us about this. He says that there are only two kinds of people in God's eyes: people who believe themselves to be blameless and those who have come to understand that they are not. In this story are two men: a Pharisee and a Tax Collector. A Pharisee was a respected Jewish religious elite who made rules for others to keep but often did not keep them himself. He was a hypocrite. The other man – the Tax Collector – was not at all respected. He was deemed a turncoat – a Jewish brother who had sold out to the Romans, paid to extract exorbitant taxes from his very own neighbors.

"The Pharisee stood by himself and prayed, 'God, I thank you that I am not like the other men – swindlers, evildoers, adulterers – or even like this Tax Collector. I fast twice a week and pay tithes of all that I acquire.'

But the Tax Collector just stood at a distance, unwilling even to lift his eyes to heaven. Instead, he beat his breast and said, 'God, I am a bad man. Help me!' I tell you that this man, rather than the Pharisee, went home justified." (22)

The Pharisee represents a person whose life is built on a subtle lie. He justifies his behavior with a self-invented morality and conveniently deems himself blameless and others faulty. His goal is not to be worthy in God's eyes, but merely to appear "more worthy than the next person" in his own eyes. Because his reference for right and wrong is transitory and subjective, he is capable of any kind of self-gratifying behavior.

When I read this, I recognized this person. I had been the Pharisee. I had counted on my own morality to make me look good in the end.

But the other person, represented by the Tax Collector, had come to know the truth: there is no wiggle room. He understood that God's standards are not attainable no matter how satisfactory he imagined his merits to be.

The Tax Collector understood that he had absolutely nothing he could bring to his trial to prove his innocence.

The Thread of Faith

Growing up, I had heard people talk about "believing in God" and "having faith," and for some reason I got the impression that believing God *exists* is what it means to have faith. But I learned that even evil spirits know that God exists. (23) Even they possess *that* level of faith. It does not impress the Creator that we merely acknowledge his existence, especially since he says it's really obvious. (24) Instead, I discovered that he asks us to go light years beyond that and believe what he *says*. He wants us to trust his words. Jesus explained, "If you abide in my word, you truly are my disciples, and you will know the truth and the truth will set you free." (25) What are we freed from? Lies. We are freed from being imprisoned by the web of untruth that evil seeks to weave around us daily. Lies are an insidious barrier that prevents us from doing what we were created to do.

I found in my research in the scriptures that trusting what God says is the sole determiner of who walks in the light and who stumbles in the darkness. This thread of trust, resulting in action, is woven throughout, comprising the very the fabric of faith.

Perhaps you were aware that John the Baptist – the prophet who prepared Israel for the coming of Jesus – and Jesus himself, were relatives? John's mother Elizabeth and Jesus's mother Mary were cousins and the two carried John and Jesus at the same time. But there were big differences. Unlike Mary, Elizabeth was elderly. (27) So, when the angel appeared to Elizabeth's husband Zechariah to announce her pregnancy, Zechariah did not believe it possible. "How can I be sure of this?" he questioned the angel. "I am an old man and my wife is well along in years." For his doubting, Zechariah lost his ability to speak.

It was restored only when he reconsidered and obeyed the angel by naming the child John. (28) In contrast, Mary immediately believed the heavenly messenger who told her that she would bear a son, despite the inconvenient reality that she had never been with a man. She was blessed beyond measure for her unwavering trust in what God told her, even though it made no sense on the human level. (29)

Then there was Abraham. Despite his shortcomings and weaknesses, he has always been held up as the model of faith down through the centuries. Abraham was the venerated father of the Jewish nation. Why? He was asked by God to leave his home and his people and set out on foot for a place not revealed to him until he arrived. (30) He and his wife Sarah, though elderly, did so without questioning. The destination turned out to be Canaan, a land that would be given to their descendants as their homeland.

When Abraham arrived, God told him that his offspring someday would inhabit the very spot where he was standing. (31) On the surface, it made no sense. Abraham had no children and his wife was way beyond child bearing age. But despite that, Abraham was informed by God that "a son coming from your own body will be your heir." (32) Then he was told, "Look up at the heavens and count the stars – if indeed you can count them all. So shall your offspring be." (33) Then it is written, "Abraham believed the Lord and the Lord credited it to him as righteousness." (34) In other words, because of his demonstrated faith in what he had said, God promised to overlook Abraham's transgressions... permanently.

Forgiveness is conveyed where there is demonstrated faith.

Years later, Moses was leading Abraham's descendants through the desert toward that very land of Canaan. Some of the people began to grumble and rebel despite being well cared for. As a discipline, God allowed poisonous snakes into their midst. He told Moses, "Make a bronze serpent, and set it up on a pole and it shall be that everyone who is bitten, if he looks at it, he shall live." (35) So Moses made the serpent and put it on a pole and advised the people to look at it and they would survive, if bitten. If they looked, they survived. Those who refused to be bothered, succumbed to the poison.

This incident would later be cited by Jesus himself as an example of the pivotal role that faith in him would play. "As Moses lifted up the bronze serpent in the wilderness," Jesus explained, "even so must the Son of Man be lifted up, so that whoever believes in Him should not perish but have eternal life." (36) These words were spoken to Nicodemus, a prominent Jewish teacher who did not understand who Jesus was. Nicodemus understood that Jesus was a popular teacher of the people. He did not realize, though he may have suspected, that Jesus was the Jews' long-awaited "Messiah" – the promised one who would save them. Many of the Jews thought that the Messiah was going to save them from their Roman oppressors but they were about to learn that it was going to be a much bigger save than that. His role would be to save them from themselves, through faith in what he would do.

And you have no doubt heard the story of the first humans and how they were instructed about the trees in the garden – specifically the tree of the knowledge of good and evil – "for when you eat of it, you will certainly die." Eating from it would lead not to physical death, but spiritual death – a separation from their Creator. (38) The instructions could not have been more basic or less restrictive. In fact, there was only one restriction and it only applied to one tree. So, the first thing that the evil one did was to attack God's words: "Did God *really* say you must not eat from any tree in the garden?" (39)

Satan had just done what earned him Jesus' descriptor, "the father of lies." (40) He hung out a carrot, suggesting that if the humans would have faith in his words instead of God's, the result would be enlightenment. "Your eyes will be opened and you will be like God,"(41) he promised. How many times has the evil one repeated this lie through the ages? How many times have you and I believed it? The first time I read this, it reminded me of how I had done exactly that, when I filled my own spiritual void with my enlightened self.

These examples of faith illustrate that God gave us a priceless gift that you seldom hear anyone credit him for: the unrestricted freedom to trust what he says – or not. Curiously, many people blame him for being *too* restrictive. And many of the same people simultaneously blame God for allowing bad things to happen on this earth, as if they would actually prefer him to be *more* restrictive. W hich is it? No one I know would trade their free will for a scripted existence - certainly not Truman.

But take a moment and bask in the brilliance: believing what God says is a process so simple that anyone can step into it, no matter what's in their past, no matter their position in life. There is no waiting period. It costs no money. It carries no requirement to memorize a complicated theology, go door-to-door to make converts or to follow a prescribed punch-list. One does not even have to be able to read or write. The opportunity to trust what God says is available to the elite, to those of no status and everyone in between. It's not complicated. The fabric of faith is simple ...and free.

I recalled the words of the angel who proclaimed Jesus' birth, "I bring you good news of great joy that will be for *all* people." (26)

So, what exactly is keeping "all people" from taking the offer?

Understanding Overcoming

I believe what keeps most of us from trusting God is that we sort of like playing the role ourselves. "Disregard God so you can be your own god." It's the original lie repackaged; the quintessential "words without knowledge." Every day we are strongly and repetitively encouraged to do this by our culture and its loyal partner, the media.

As you know, the ploy worked famously and still does. The inevitable separation affected not only the first humans, but all humans thereafter, including you and me. We've all inherited their selfish tendencies and believed the lie that isolation from God will result in greater happiness.

In practice, it alienates us from greater happiness.

God refers to separation from him as "death." What does he mean? You may remember hearing the parable that Jesus told us about the prodigal son (the word "prodigal" coming from the Greek word for "unrestrained.") In the story we are introduced to a boy who believed a destructive lie. Though we are not told precisely what that lie was, we can assume that he thought he knew better than his father and wanted to be rid of him so he could exploit his freedom. Indeed, he was unrestrained - self-centered in the extreme. He even requested his inheritance while his father was still alive, the ultimate insult. In effect, he was saying, "I don't have the time to wait around until you die, Dad. Give me my inheritance now."

But instead of withholding it, his father freely gave it to his son. He did not try stop him. The son departed his home and it was not long before he had wasted all his money on selfish, ill-advised pleasures. As a result, he had to take an insulting job that paid peanuts. Without warning, the economy turned bad and there was not enough food to sustain him. When he saw that he was going to die, the boy began to realize what he had done to his father, his family and to himself. He had thrown away his generous inheritance, but more importantly, he had discarded the precious relationship he had with his own father.

No longer considering himself worthy to be called a son, he started back home to see if his dad would just hire him as a worker so he would at least have something to eat. But his father saw him coming at a distance and ran to meet him with open arms, exclaiming to all, "My son was dead, and is now alive again!" (42)

Why does the father say that his son was dead?

God is warning us sternly that believing lies about him ultimately will lead to our spiritual death. The son's relationship with his father was nonexistent. He had traded a relationship of inestimable value for perishable trinkets. We all have seen elements like this on our own journeys. And like that boy, we first have to become convinced that we lack the resources to run our own game indefinitely. We must turn from our destructive road. It is critical, because left untreated, disbelieving God results in this most foreboding kind of death – the one that is unending. Jesus called it the "second death."

I looked throughout the scriptures to find something to help me find a way around such an outcome. I had learned in my Catholic upbringing that after my natural death, my soul would arrive at a place known as "Purgatory" for rehab before I could apply to be admitted to heaven. But I am sorry to report that there isn't any such place referenced anywhere.

Instead, we are warned that the second death – the fate of those who disregard God - will be very real, very awful and very permanent. It was compared by Jesus to "Gehenna," the landfill just outside Jerusalem where residents burned garbage. (43) That foul place was even littered with the rotting corpses of criminals the Romans had discarded like trash after execution. In Gehenna, the fires were never extinguished, smoldering day and night. Awful as it may seem, the most terrifying part is that it is open-ended. Infinitely worse than a physical death sentence, the second death is a life sentence in a place of never-ending pain and destruction.

If that sounds awful and frightening, I believe it is meant to. In fact, I believe it's meant to scare the hell out of us. I believe we are told about it so we will seek out a way to avoid it.

But my mind began objecting, "A good God would not allow such an awful thing to happen to the people he created and loves." Yet, this very question on my part, showed my profound misunderstanding of God. Why? First, because it's plainly impossible for you or me to evaluate the methods and motives of the Creator of the universe, vetting his suitability as we would a subordinate. Remember that he himself asked, "Where were you when I laid the foundations of the earth?" Our view of reality is so restricted by space and time that though we may ask such questions, to think that we could answer them is ludicrous. Imagine you and me thinking we can determine if God can measure up to our standard of "good."

Second, like the wise father in the parable, God is actually all about free will, not manipulation. Our future is our business. He advises us, but he never forces us to believe him. He watches for us, but he always lets us make the calls. If we choose rebellion, he lets us continue down that road in the hope that we will learn. If we turn around, he meets us before we even see home.

He was already planning his reunion with us right from the beginning. Read this from Genesis:

God declared to the evil one: "I will put enmity (bad blood) between you and the woman, and between your offspring and hers. He (Jesus) will *crush your head*, and you (Satan) will *strike his heel*." (44)

This prediction says that in future generations there would come someone born of the first woman's descendants who would be attacked by evil, but who would not be overcome by its poison. Instead, he would have the power to deal it a fatal blow. This person would have to be a man who was more than just a man. He would have to be a man who was also God.

The Jews understood that this person would be the long awaited "Messiah," about whom so much had been written. There were hundreds of specific prophesies that would foretell this Messiah's life and death in the scriptures, all written hundreds of years before his birth. Why? So he would be recognized by some when he came, and so the prophesies could be shown to be accurate predictions issued by God himself.

The scriptures that foretold the Messiah's characteristics were amazingly specific. His mother would be a woman who had never had relations with a man. (45) (That alone would have positively identified him.) But add to that, it was written that he would be a descendant of the Jewish King David (46) and would be born in a tiny town named Bethlehem, whose name meant the "City of David." (47) He would be misunderstood by his own people, rejected and abandoned. (48) He would be put to death. Then, he would return to life within three days. (49) Old Testament prophets under God's direction wrote down these and hundreds of other specific details about the life of the Messiah. They were fulfilled, one by one, in the person of Jesus, and documented in writing. It's obvious that no other person who has ever lived could have met every one of these criteria.

As predicted, few Jews recognized that Jesus was their long-awaited savior, but his disciples would later recall his words and connect them with the ancient prophesies. The simple faith that this group of 12 people had in his words, and the faith of the millions that would come after them, would comprise the very weapon that God would use to land that crushing blow.

But if faith is a good thing, how could it also be a weapon?

For dictators, cultists and political leaders who practice evil, faith in the words that God speaks is absolutely a threat. It is said that "truth is the first casualty of a dictator" and they especially don't want their subjects obeying an authority higher than their own. That is why those who speak the truth in authoritarian regimes are so often marginalized, oppressed, imprisoned or killed. Examples of authoritarians that murder, in cold blood, people who speak the truth are too numerous to mention. No wonder faith in Jesus - the truth incarnate - is disallowed by such regimes.

If truth can be that dangerous, it explains a lot. It explains why most of the disciples were exiled or executed by the Roman government when they would not recant what they knew to be true. Their faith in what they had personally witnessed was deadly to those in charge, whose "words without knowledge" – today referred to as "narratives" – asserted that *they* were the deities of their day.

But the stubborn faith of the earliest believers – and their personal knowledge of what they had seen with their own eyes – could not be imprisoned, tortured or executed out of them. The Jews tried. The Romans tried. But despite the severe oppression, the tiny band of frightened friends and courageous fishermen knew what they had witnessed, and with God's help, they boldly stated it for the world to hear; for you and me to hear.

A sincere trust in God's words is indeed the most potent of weapons. It fought its way from the stillness of a tomb, past armed soldiers and into the very throne room of world power. Incredibly, within just a few hundred years of Jesus' crucifixion, the Roman Emperor Constantine was himself baptized. Then he personally put an end to the persecution of Christians, making way for the faith to flourish. The truth about God was trumpeted from the highest levels of the Roman Empire – ironically, the very one responsible for Jesus' execution.

The effect was to cause God's words to metastasize.

Indeed, the word that God speaks is commonly symbolized as a sword in the scriptures. And since Jesus himself is the embodiment of God's word, his role is to be the ultimate divider of good from evil, both on Earth and eternally. Trusting him will bring forgiveness and peace to the faithful; disregard for him will cause the unfaithful to stumble headlong into Gehenna.

But you must understand, at your and my level, God's truth is the only weapon that can serve as an antidote to the evil that resides within us. From the book of Hebrews: "For the word of God is alive and active. Sharper than any double-edged sword, it penetrates even to the dividing of soul and spirit, joints and marrow; it judges the thoughts and attitudes of the heart." (50) Believing God exposes the darkness that we harbor and allows his Spirit to change us from within, becoming a "light for our path." (51).

Jesus proclaimed, "In this world you will have trouble; but take heart – I have overcome the world." (52) When I first read this amazing statement, frankly, I did not understand it. So, I did some research and here is what I found:

"In this world." The word "this" jumps right off the page. He is telling us that *this* world we currently live in is not the *only* world, and this world is not the ultimate one. There is a better world that he refers to; a place so incredible that we don't have the capacity to imagine it.

"You will have trouble." No one escapes it. Jesus states that we *will* have trouble in the current world. It's not a "maybe." If you have lived any length of time and are honest, you have to admit that this world is a wreck. It is littered end-to-end with unsolvable problems at every level. There is a swamp of relationship chaos we all must swim through every day. Perhaps you believe that you have conquered your troubles? Perhaps you have moved on? I invite you to ask someone who knows you well if they agree that your relationship with them is trouble-free.

"But take heart." Here is encouragement...

"I have overcome." Notice he did not say that he would teach us how *we* can overcome trouble. He is not describing a self-help program. Instead, he is stating unequivocally that he himself is the solution. He does not promise to accomplish this someday in the future. He says,

"I *have* overcome the world." (52) It's past tense. He elaborates, "no one comes to the Father except *through me*." (53) Jesus is asserting – with no mistaking – that he personally is our only solution.

In the book of John, he makes it clear that we can join him in overcoming by simply believing he is who he says he is and acting on it. John tells us, "And this is the victory that has overcome the world – our faith. Who is it that overcomes the world? Only the one who believes that Jesus is the son of God." (54) The Result: "The word of God lives in you and you have overcome the evil one." (55) Through the act of believing what God says, his Spirit joins with ours, providing the power to vanquish the lies, the selfish thoughts and the bad habits that have caused us to stumble in the darkness.

"... The world." Notice the repetition of this phrase. When the Jews repeated a phrase in speaking or writing, it was their emphasis point, I learned. The very "world" that attacks the truth is the same one Jesus came to Earth to overcome for you and me.

It is impossible for us to overcome the evil of an entire world when we can't even extinguish that which resides within our own selves. How can you and I – imperfect as we are – perfect anything? Incredible as it seems, the scriptures say that God volunteered *himself* to be sacrificed on our behalf. He states that he has assumed the blame for all of the world's evil deeds, big and small; yours and mine. Jesus allowed himself to be executed. (56, 57) It's unthinkable. This selfless act is what sets the true, loving God apart from imitators.

Sacrifice was not a new concept to the Jews. Long before Jesus' time, the Hebrew nation was commanded to sacrifice spotless lambs and trust that the lamb's fate should rightly have been theirs. If they trusted in this, they were absolved of their transgressions. We now understand that the substitutional death of the lamb was just a picture of what God would eventually do *to himself*. He did it because love and self-sacrifice are his core nature. No wonder we can't evaluate God's goodness. It infinitely exceeds any standard we would invent. We cannot imagine letting ourselves be tortured on behalf of someone else for any reason. But imagine an innocent person volunteering for a criminal's death to spare someone he or she had never even met?

But unlike the lambs of old, Jesus would subsequently return to life, proving to all that he was not merely a man, and inviting you and me to follow in his footsteps - to overcome.

Since the Creator of the universes is, by definition, all-powerful, Jesus could not have been murdered. Who would be so powerful as to kill one who is all powerful? So, if God himself was crucified like a despicable criminal, it means that he absolutely allowed it. And it also means that it had to have been for a very important reason. Indeed, you and I must be very valuable to our father. As Jesus explained to his disciples, "Greater love has no one than this: To lay down one's life for one's friends." (58) This is the clearest of demonstrations that his ways are not our ways. He paid our ransom with his very own life.

Jesus had told his disciples ten different times what he was about to do. He even explained to them that no one was going to take his life away from him: "I lay it down of my own accord," he stated. "I have authority to lay it down and the authority to take it up again."

Indeed, we have no defense against the charges. But God himself provided our exoneration. He made himself available to stand accused in our place. All we have to do is to trust him. That's only way out of the burning house we call home.

Rising Certainty

There are some who doubt that Jesus' torturous death and inexplicable resurrection actually happened. This should not surprise us. Despite no evidence to the contrary, there are those who think that astronauts did not really land on the moon in 1969 when it happened in full view of the entire world, live on television. Despite overwhelming evidence, people can – and do – reach conflicting conclusions.

Certainly, there are documented cases where someone who was pronounced dead was later determined to be alive because they were never actually dead. But that is not what we are talking about here. Jesus was impaled and hung up for six hours, stabbed so he would bleed out, and was acknowledged to be dead by both his executioners and his own family. Then his body was wrapped and buried in a sealed tomb.

There was no doubt that he was dead. It happened in full view of the public. Neither Jesus' enemies - the Jewish authorities - nor the Romans who executed him, ever claimed that he had survived the crucifixion. The Jewish leaders acknowledged that he was dead, and, of course, the Romans would never have removed a live body from a cross of execution.

Yet, it certainly would have been advantageous to the Jewish narrative if they could have claimed that Jesus had survived, because what happened next caused them a real problem: they had no talking points to explain his disappearance from the guarded tomb or his subsequent reappearance in public - alive. And the stickiest part was that Jesus himself had stated publicly – in advance – that it would all happen as it did. It was a public relations nightmare. And in time, the people would realize that the nation of Israel had succeeded in executing its long-awaited Messiah solely because he dared to speak with an air of authority that diminished the religious elites.

My study of Christianity was introducing me to a Creator who cared so deeply about his creation that he was willing to personally die in its place. This God is unique in all the universe for that, and also for what happened next. You see, in all of recorded history, no man has ever shown up alive after his own execution. When Jesus performed *that* feat, it became obvious that he was no mere man.

But it is critical for us to be absolutely certain that his death and resurrection really happened, because if they didn't, the central fixture of the Christian faith is in question and so is the very integrity of God. These events are presented as historical events, not as parables. So, wanting to be sure, I did some research to see how certain I could be that Jesus actually did come back to life.

Easy to Disprove. The first thing I learned was that there is no evidence that has ever been unearthed that Jesus *didn't* rise from the dead. Of course, the best evidence would have been the tomb with Jesus' dead body still in it. But that apparently did not exist or it would have been trumpeted loudly by the Jewish authorities. The case would have been closed. However, once buried, no one ever claimed to have found him dead, anywhere. Instead, three days after his execution, his tomb was acknowledged by both Jews and Romans to have been inexplicably opened and inexplicably empty. Despite possessing all the levers of power, even the Romans and their legions could not secure the tomb and – once it was compromised - could not locate a dead Jesus anywhere? Both should have been the easiest of tasks.

Predicted In Advance. As mentioned before, it had been predicted for centuries – in writing - that the coming Messiah would die and rise again. Confirming details about the crucifixion method were even recorded hundreds of years in advance. For example, in Psalm 34, (85) consider the prophesy that his bones would not be broken. Unlike the standard victim of crucifixion, Jesus' legs were intact after his execution. Why? Because breaking a man's legs only served to hasten death and Jesus was already dead hours before the end of the day after he declared, "It is finished" and immediately died. (86) [It was unheard of for a victim to proclaim his last breath. And because those being crucified typically lingered, Jesus' announcing the conclusion to his own execution caused the Roman guards to remark that he could not have been a mere mortal. It is recorded that they exclaimed, "Truly this man was the Son of God."(59)]

Jesus had told his disciples many times what he was about to do. Though predictions are not proof, this writer believes that you could look through all of human history and never find another person who predicted publicly that he would be executed and subsequently was able to convince billions that he had come back to life. The fact that no one has ever made such a claim and then pulled it off is de facto proof that it happened.

Eyewitnesses. A modern court of law recognizes eyewitness testimony by multiple persons as the most convincing means of establishing what is fact. The New Testament is replete with the personal written statements of three contemporary witnesses – the disciples John, Matthew and Peter [whose testimony was transcribed by Mark] – each of whom saw or knew of Jesus' death firsthand. The same men later stated publicly that they met with him multiple times after he arose across a span of more than a month. We are also told that not only these, but as many as 500 others, some who were strangers and some who Jesus knew personally, saw him and met with him on dozens of occasions during that time. Their eyewitness testimonies would today be considered incontrovertible legal proof that Jesus' crucifixion and resurrection occurred.

Refusal to Recant. As noted before, Jesus' disciples and associates chose martyrdom, imprisonment, torture and exile rather than say, when challenged, that Jesus' resurrection did not happen. If the reality of the resurrection were in doubt, certainly some of them would have flipped on their testimonies before offering up their freedom or their lives. But there is not one record that any did. Instead, history records that every one of the disciples suffered either exile, crucifixion, beheading, hanging, bludgeoning, stabbing or stoning, and each one of them steadfastly stuck to their story.

Contemporary Readers. Because the first writings attesting to his rising were circulated while the eyewitnesses to the events were still alive, they had to have been accurate. Had they been fraudulent claims, they would never have been approved by the very readers who would have been in a position to know better.

The Guard Guarantee. The Jewish leaders did themselves no favors when they insisted on Jesus' public tomb being watched 24-7 by a contingent of heavily-armed Roman soldiers. Thinking themselves clever by preventing sympathizers from stealing Jesus' body so they could claim that he had risen [a claim which would never have been believed without Jesus returning to life to prove it], they inadvertently provided proof of the resurrection themselves when they insisted on the guard. Obviously, no one could have overpowered a contingent of armed soldiers, broken the official seal, moved a stone weighing tons, entered the tomb and gotten out with the body, totally unnoticed. But that is the whopper the Jewish elite circulated after their carefully guarded tomb went empty right on schedule – the third day after the crucifixion.

Interestingly, there are some who have speculated that the Jewish authorities were never really concerned about the disciples stealing Jesus' body. After all, they were just eleven men – a scattered handful of frightened fishermen – who had run for their lives. Their leader had just been executed. They had no serious weaponry to fight with. Why would they hatch a plan guaranteed to get them killed as well?

It has been proposed that the real reason the Jews wanted the tomb guarded was, given his many miracles, some may have worried that Jesus might actually come back to life. They may have imagined that the guards would provide insurance that Jesus would remain in the grave. Is it possible that some of the very men who sent Jesus to his death feared he would not be bound by it?

The Cultural Reboot. The final proof is the one you and I can attest to: the fact that we even know about the resurrection today makes it a certainty that it happened. Why? Consider for a moment what you personally would need to accomplish in your lifetime for your name to be remembered as a household word more than 2000 years from now. And what would it take for your every word and action to continue to be discussed in detail two millennia after you lived? You would need to perform some amazing feat that would never be forgotten.

I believe that you would have to pull off something considered to be physically impossible ... like ... rising from the dead.

It is undeniable that *something* earth-changing happened with Jesus 2000 years ago. He was not just some man executed for having revolutionary teachings. Why would people still talk about him? Such rebels were regularly put to death by Rome. Jesus would have been just one more. But to die such a death and then return to life in full view of all, having stated publicly that you would do it? Yes, that is the kind of thing that could reboot an entire civilization. It was – and still is – unheard of, which explains why we still talk about what happened in the year 33 A.D.

Faith in the certainty of the resurrection is what gave birth to the Christian Church and formed its foundation. The news spread from Jerusalem to the rest of the world on the very roads of the Roman Empire. Its unprecedented growth occurred despite perpetual persecution – an obstacle that has had to be overcome daily since its inception. The fact that the Church has thrived most vigorously where it has been oppressed most vigorously seems counterintuitive, but it really isn't. The faith began with the persecution of Jesus, who invited his followers to "pick up your cross and follow me." (60) He was telling those of us who trust him that it would always be that way.

But understand that the very idea that a man died and came back to life is a very, very heavy lift. No rational person would believe such a story unless a) they had personally seen the person die and b) the victim later stood bodily before them – alive. It would take both for a rational person to consider that a resurrection had truly taken place. Imagine if you had attended a friend's funeral only to be told a few days later that he had returned to life. There is no way you would believe such a story without first going to see your friend standing before you, alive.

Not surprisingly, among the eleven disciples, even some of *them* were doubtful.... until they saw the proof. The disciple Thomas was convinced only after personally seeing Jesus in the flesh, the scriptures say. (61) All of his doubt was apparently obliterated because Thomas spent the rest of his life telling others about Jesus' return to life before being martyred in India, refusing to change his story.

Since the very foundation of the Christian faith is based on Jesus' bodily resurrection, then a) and b) had to have occurred. It could not have been anchored on people merely "saying" that Jesus rose, without indisputable proof. The first believers would have insisted on seeing that proof for themselves or Christianity would never have become a movement.

Had it been lie, the story of Jesus' resurrection would have been labeled as such and would have died soon after its founder.

Everyone Born Of The Spirit

But the movement obviously did not die; instead, the news of Jesus' triumph over the grave proceeded to recalibrate civilization.

_ The document that describes and chronicles the faith, which came to be known as the *Bible*, became the very first book to be printed on a press and continues to be the most widely circulated book in Earth history.

_ The foundations of modern morality and our system of laws were established upon its Ten Commandments.

_ Even the worldwide reference for the numbering of the years was reset to Jesus' birth year – "Ø A.D." ("A.D.," of course, standing for "After Divinity," a tacit acknowledgement that in "Ø A.D." something extraordinary took place that was divine.)

It was also clear among those very first believers that something unprecedented had occurred at the personal level. Jesus had told them that he would send them a helper (62, 75), – his Holy Spirit – which would take up residence *within* them. It first happened at an event that came to be known as "Pentecost," when the once-cowardly disciple Peter was given the confidence to boldly explain to the masses about Jesus' death and resurrection and what it would mean for them. (63) It was recorded that upwards of 3000 people believed Peter's words that day. Pentecost kick-started the Christian Church, and from that day forward, all who would trust in Jesus's sacrifice for them would be reborn as "new creations." (64) The change was, and is, profound. God's

Spirit began confirming with ours that he knows us. The father has met us halfway. The estrangement has ended. In fact, Jesus actually stated that those who trust him, he regards as *his friends.* (65) How amazing that instead of requiring perfection of us, instead he meets us right where we are on our journeys and befriends us through his Spirit.

In another place, Jesus calls out directly to you and me: "Here I am! I stand at the door and knock. If anyone hears my voice and opens the door, I will come in and eat with that person and them with me." (66) The act of dining in the ancient times indicated a level of intimacy and trust. Jesus' use of that metaphor tells us that the spiritual transformation that takes place is not merely academic – like ascribing to a written creed or set of rules. That's because we are trusting not in a philosophy, but in a living person with whom we can share friendship.

Indeed, he is one with whom we can dine.

God's Spirit also gives us a direct means of communicating with him. Unlike earlier times when priests were the required "go-betweens," his Spirit now makes it possible for us to speak to him one-on-one, and for him to speak to us. (67) He plays the role of an intercessor, since we often don't know what to say when we pray. (68)

Until that point in my life, I was like a person sent on a grocery errand to buy food for a stranger. There were lots of great choices on the shelves that I liked, but I was never sure what I should select because I didn't know the sender. But when I believed God's words, his Spirit began advising. Before I had been hesitant, unsteady and ambivalent, but with his presence came light. I was better able to make life choices confidently. (96)

Recall the Pharisee Nicodemus who was interested in learning more about Jesus' teachings? One night after dark when he thought he would not be observed, he requested a personal meeting. He told Jesus straight out that it was obvious to him he was from God. We will never know where Nicodemus was going with his comments because Jesus redirected him, "I tell you the truth, no one can see the kingdom of God unless he is born again." (69) Nicodemus had been taught from youth that following the laws of Moses and the doctrines of the Pharisees provided the way to earn one's eternal place with God. But to his surprise, Jesus took a sharp turn, telling him instead about a new "birth" that was first required. Jesus marveled, "You are a teacher of Israel and you do not know these things?" (37)

He said that no matter how hard we work, we cannot earn God's forgiveness - and a future with him – even if we work in churches, build homes for the homeless or give all of our money to the poor. He stated, "Many will say to me on that [last] day, Lord, Lord, did we not prophesy in your name and drive out demons in your name and perform many miracles? Then I will tell them plainly, 'I never knew you. Away from me, you evil doers.'" (70) These petitioners would not gain the most valuable inheritance – an unending friendship with God – because they made the mistake of trying to prove themselves worthy solely by their actions. God is addressing people just like you and me who lived their entire lives and simply decided not to believe him. Then they proved they did not believe him by their deeds.

"I never knew you ... Away from me ...." When I read those words, frankly, they shocked me. They shouldn't have. In day-to day-life, it's not so much *what* you know but *who* you know that determines if you rise or fall. It should not have surprised me that it's the same in the spiritual realm. Knowing *about* God is not knowing God.

Like most of us, the respected teacher Nicodemus did not understand what a new birth of his spirit would mean, responding, "How can someone be born when they are old?" (71) Jesus used the occasion of Nicodemus' honest question to explain to him gently, "The wind blows wherever it pleases. You hear its sound, but you cannot tell where it comes from or where it is going. So it is with everyone born of the Spirit."

From all accounts, Nicodemus trusted Jesus, and to use his very words, was "born when he was old." Despite being one of the Pharisees, he boldly volunteered to assist at Jesus' burial.

When we begin to trust what God says, he causes our spirits to be reborn. It's a permanent change. We are told that we become God's forever and no one is powerful enough to "snatch us out of his hand." (72) We gain the power to turn from the destructive road to the constructive one. We gain certainty about our ultimate destination, but as in Jesus' metaphor, living moment to moment becomes an adventure based on faith and trust.

Speaking of our ultimate destination, my early church understanding was that life with God in heaven would be a tedious existence – a rather boring time that would drone on endlessly with harps, etc. Why would I look forward to that? But that picture must have been based on more "words without knowledge" because even a short a survey of what God tells us about a future with him does not match that.

Instead, what lies ahead for those who trust him is beyond all description – literally. The Apostle Paul tells us, "No eye has seen, no ear has heard, and no human mind has conceived of the things God has prepared for those who love him." (73) This kaleidoscope of creation that surrounds us on Earth is so vast and so varied that we can perceive but a fraction of it. But the Creator of universes tells us that there is an infinitely higher level that is so stunning that there are no words that can describe it. Jesus just says that if he knows us, we are invited. "I am going to prepare a place for you so that where I am, you can be also." (74)

I decided I should seek that invitation. I wanted to see for myself. Spending forever with a friend like that was worth any sacrifice I needed to make.

Road Test

I had precious little to show for my six years in the Quad Cities. It was at that point when I finally had enough.

Things didn't look encouraging – an understatement. My brief broadcasting career had plateaued and I was sliding backwards. It was apparent that I needed to change professions but I didn't have the resources or the time to return to school. I had made few friends during my time along the Mississippi and I had not been dating much. In hindsight, it was good that no one was depending on me right then. At the six-year mark I was like a parachute jumper who just realized that his chute was not going to open. I had limited options and very little time.

It was at this critical juncture I was advised, "I've got this. Keep going."

12 Months To Go. The United States was in a deep recession and inflation had been in double digits. It was hardly the time to begin a business. Even so, I had continued to investigate that "radio in a campground" idea I had been pondering around the campfire.

I made some calls and found out there was one such radio system installed in Iowa only 5 miles from my house. Thanks to the kindness of the Iowa Department of Transportation, I was able to identify the manufacturer of the transmitter, which turned out to be just a few hours away in Minneapolis. My radio engineer friend offered to assist, so I emptied much of my savings to purchase a set of the company's equipment. Unfortunately, I still lived in the tiny apartment at Charlotte and Carey Streets. There was no backyard where I could experiment.

So, I tried an approach that was very unlikely to play out. I did a cold call. I drove up the river to a state park I knew of and knocked on the front door of the supervisor's home. I had to do a lot of explaining, but it turned out that he was agreeable to letting me set up the radio system in his park for a season.

I learned a lot in the experiment, but there was apparently a lot more to learn because the signal did not cover the way I had anticipated. But despite that, the Minneapolis company needed a technician for their next project and (sadly) I was the closest thing they could find. They sent me to San Antonio to install one of their systems at the airport there. I took the week that followed Thanksgiving to make the drive down and back and told absolutely no one where I was going. I knew my boss would not have been happy I was moonlighting.

One moment during that visit stands out vividly: a worker was using an asphalt saw to cut grooves into the pavement for ground wires. Suddenly the saw blade shattered and shot slivers in all directions. They were flying up in the air and landing all over the parking lot. Someone could easily have been badly hurt. In fact, one of the shards would have cut right through me had it not been stopped by a cabinet door which I happened to be working behind. If the door had not been standing open, I absolutely would have been severely injured; the gouge in the 14-gauge steel door was an inch deep and centered on my chest. The young man running the saw examined the door and looked at me behind it, shook his head and commented, "There is absolutely no doubt that somebody wants you alive."

The installation actually went rather well despite my lack of electronics training. I also lacked the right kind of vehicle to show up in. [Picture a Honda Accord with an extension ladder strapped to the top and bungee cords front and back.] Yet, incredibly, the airport and the Minneapolis company seemed satisfied, so I decided to see if I could sell their products. I even started calling myself a "company."

9 Months To Go. The day before the 4th of July, I got a call from the print shop to say that my brochures were done. They featured the new company name and logo. All I needed now: people to buy the products and money to run the company.

8 Months To Go. Then I got another call from the Minneapolis firm. They told me about a new project in Seattle. But when I returned from that trip, they announced that it would be the last one because they had decided to stop manufacturing transmitters. Sales were not satisfactory. They handed me some leftover sales leads and said, "Good luck. It's been good working with you." Suddenly I had no product to sell.

Yet, I continued to be advised, "I've got this. Keep going."

7 Months To Go. On Friday, September 9th, it would be only 7 months until I would lap my seventh year in Iowa. That is when I heard the sound of the studio door opening behind me. I looked around to see my friend, the radio engineer. He wasn't smiling. He had been directed to usher me out of the building at the end of my shift, take my key and lock the door behind me. I was terminated. The radio station was in red ink and they had decided to let the highest paid staff go.

If you are keeping track: I was without money, customers, a product to sell, and now, a job to pay the bills. Finding a new job would not be easy with the word "fired" next to my most recent employment.

The following Monday I was downtown at the Moline unemployment office before the sun was up. The autumn breeze swirled old sheets of newsprint around in the dark alcoves between the buildings. Indeed, my broadcasting career was already old news. The national unemployment rate was 9%. The line was long.

To get an unemployment check back in that day, you had to report where you were trying to find work. They didn't want to hear that you planned to start a business and become self-employed because nine out of ten new businesses fail. So, I made a half-hearted effort at job hunting to get the checks coming.

Was I going to take another job in broadcasting? The industry would pay a lot more than the convenience store down the block. I did take a few calls from associates who advised me about openings, but something felt wrong. Something told me that the time to retrace my steps was over. I really didn't understand much about this new Spirit. I only knew that he was providing guidance. But soon it would provide even more.

6 Months To Go. Winter was waiting in the wings. In the park near my apartment stood a massive white oak tree that watched over the Mississippi from its prominent perch on the bluff. Its towering limbs waved their warning: the seasons were about to change. Across the river, the amber brick buildings at the Rock Island Arsenal stared back as they had for more than a century. The grizzled old tree probably recalled the days when those structures were built. Beneath its leafy canopy was a bench with an expansive view of it all. It was there that I would go and think.

I asked God what I should do. Broadcasting? Self-employment? More education? I knew the answer before the question even left my mind. I was to pursue my fledgling business idea despite having no luxuries like a business education. Did I mention that I was living in a house soon to be condemned? Of course, I had no business plan and no line of credit. I didn't even know what a line of credit was.

Even with all that, I continued to be advised: "I've got this. Keep going."

Trust me, I was getting little encouragement from family and friends. I even found out my ex-girlfriend was telling people that I had "given up my broadcast career to become a DJ in a campground." Flattering? I didn't care. She had no lights on her street. And I was confident that I wasn't seeing green lights if I was supposed to stop. I would "take those long nights and impossible odds." Meanwhile I was watching the unemployment checks arrive, knowing they would cease when Spring came.

5 Months To Go. One day I noticed that someone had placed a classified ad in the newspaper for leaf-raking volunteers. I showed up. It turned out to be a church singles group that was helping homeowners whose yards were knee-deep in fall's remnants. I was wondering if I might meet a like-minded person there. I helped them rake for a few hours and left them my phone number in case they needed me again.

Before mobile phones there were "telephone lines" – copper wires that carried phone calls to homes and businesses. If you could not afford an office and a secretary you absolutely had to have an answering machine on your line to record messages – what we now call "voicemails." When you returned to the office you would wind back the tape and review the recordings.

One day I picked up a message from a school administrator at tiny Slippery Rock University in Western Pennsylvania. He was requesting information on radio transmitters. I had not yet found a new product that I could sell, but privately I reasoned that I could wrap in a family visit to Ohio on the way. So, I called the man back to tell him I could stop by the next time I was in the area, and "by the way, that would be next week." He happily agreed to a meeting.

It was a 9-hour drive from Davenport to Slippery Rock. That gave me time to think about how I would approach the conversation. I had decided to drive halfway across the Midwest to tell a potential client that I *didn't* have a product he could buy? So, I decided to offer to help them with planning. I would promise to reconnect when transmitters were again available.

The meeting with the University staff went surprisingly well, and when I mentioned that I was currently in between suppliers, they looked at one another and offered, "Maybe you should consider *this* one," pushing a brochure across the table. It was a product sheet from a small Pennsylvania company that had just started making precisely the kind of transmitter that the University was interested in. It was also precisely the kind of transmitter that could replace the one the Minneapolis company had just discontinued. The very next day I called the manufacturer and learned that I could become a dealer.

On the drive home, I recall thinking, "I was not even asking God if he would provide a new product that I could sell. How exactly did this all happen?" Then I recalled the parable about birds and the grass and the flowers. 11

Slippery Rock never did buy a transmitter, but now I had a product. But, how would I advertise it? Product promotion takes real money and no bank was going to front me real dollars without real collateral or a real business plan.

4 Months To Go. Out of nowhere, the church leaf-raking guy called and asked if I would be interested in joining something called a "small group." I wasn't so sure what that was, but I stopped by his house one evening where I met another gent who had been invited also. It was explained that this was an opportunity to learn more about what was in the scriptures: no obligation, no cost, and if I wasn't interested, no problem. I didn't have to join a church. I commented that I didn't even own a Bible and the leaf raking guy was all ready for that. I went home that night with my own personal copy.

3 Months To Go. Making the business go with no seed money was going to be a formidable challenge. In those days there was no internet, no websites, no social media and no search engines. People didn't even know what a fax machine was yet. To promote a new product, you had to purchase media advertising or do a direct-mail campaign. This was a box I didn't think I would be able to punch my way out of, so I returned to my oak tree bench and said, "Now what do I do? Abraham couldn't become the father of the Jewish race without children and I can't get a business going without money."

It was as if the question was expected. Soon I had a perplexing answer. I was to proceed "as if." To prompt others to believe in the product, I was to do things "as if" I had always done them. I was to behave like the best radio expert I could be. But how does one instantly become expert at something one knows little about?

It wasn't long before the idea came into my head to start a newsletter that featured stories about the kind of radio transmitters that I was hoping to sell. I had no writing experience and hadn't sold anything that I could write *about,* so generating content for such a publication was going to be nearly impossible. But despite all that, I seized on the newsletter approach since it was the only one I could afford.

Compounding things, there were no word processing programs that would allow me to layout a newsletter on a PC. (That would be Apple's Pagemaker software, still 5 years in the future.) Nor were there any affordable printers that could reproduce quality images and text. So, I went back to my friend at the print shop to see what he could suggest.

As I suspected, typesetting a newsletter on a conventional press would be very expensive. But as I was departing, my contact took me aside and told me about a man named Hank who he thought might have some ideas.

It turned out that right down the street from my apartment this Hank owned a gas station that was no longer a gas station. He had converted it into a graphics business. In the original office area was his new toy: a big, black computer the size of a Mini Cooper. Hank showed me how he could use a keyboard to type words into the behemoth. Then he could select the font, point size and column width and – voila' – after lots of humming and clicking, out of a slot came professionally printed words in neat columns. The photographic-type paper had to dry before you could touch it, but once it did, it had all the appearances of real printed text, but for pennies on the dollar.

So, I headed back to my kitchen "office" and wrote down some simple stories about the projects in San Antonio and Seattle. At the library I mined some facts from a magazine article about a radio station at LAX airport. Hank let me enter the content into his machine and soon I was back at home with the three stories on paper. Photocopying the paste-up produced a readable newsletter. On the masthead in big letters I printed the title, "Limited Area Broadcasting Report."

I read it over. Amazingly, it did look "as if" the publisher knew something about the topic. The cost, including the paper, folding, printing and postage, was 25 cents per copy. I produced 500, so the entire campaign cost $125.00. I hand-addressed them to the contacts the Minneapolis company had given me and threw in some National Park addresses I found at the library. Soon, they were sorted and bundled with rubber bands and piled into the "bulk mail" bag at the post office.

Then, the wait began.

2 Months To Go. I had been anxiously monitoring my phone and my mailbox. Nothing. I reminded myself that bulk mail is the lowest priority and takes double-time to get where it's going. My mind started to entertain, "what am I going to do when this effort fails like most new businesses do? After all, who would actually buy sophisticated electronic equipment from an unemployed DJ working out of a dilapidated apartment in Iowa?"

After a week or so I began to get scattered phone calls that referenced receiving the mailing. Then a small order or two arrived. It was not great, but after 30 days I saw that the income for the month was about the paltry sum that unemployment insurance was paying. So, in a leap of faith, I decided to cancel my unemployment. With no small amount of trepidation, I dialed up the unemployment office and told them I wanted to stop receiving my $700-a-month.

"So, you've found a job?" returned the voice.

"Well, I *sort of* did," I responded.

Silence.

"Uh, you either found a job or you didn't," the lady said, as if she had just returned my best tennis serve.

"Well, yes, I *did* find a job," I countered.

"And where is that job?" she inquired, her tone like a kindergarten teacher's.

"It's a job that I started myself," I explained. "I started a business in my home."

Silence again. It was obvious that this civil servant didn't have a box to check that said "Subject has Started a Home Business."

"Oooh Kayyy," she said, sounding as if she would be ever so pleased to get me off the phone. "Thank you sooo much for letting me know." [Click]

She just hung up.

I looked out the window and wondered if I had just sealed my fate.

It had been a particularly frigid winter along the big river and now that the ice was melting, you could hear engines revving up. Barge traffic had started to chug through the locks again. There had been times I would wonder what the heck I was doing. With so little income, I had to jam my expenses to the floor. I even hung plastic sheets in the apartment's interior doorways and rolled a space heater room-to-room to avoid running the central furnace. But I did not care. I was off unemployment. I know it sounds like I was very, very poor but I was also very, very self-employed and no one was going to take that away. One afternoon I walked down to the corner drugstore and purchased a book titled, *How to Start a Home Business.* It was the perfect title because that is exactly what I had.

That and my faith were *all* I had.

1 Month To Go. The sun rose a little higher in the sky every day. The pin oak in the neighbor's yard had jettisoned its leaves and they lay scattered, tempting the wind to whip them onto the broken bricks that carpeted the corner of Charlotte and Carey Streets.

With improving weather, I had taken to walking a daily circuit downtown and back that included the post office, the library, the print shop and the bank. The routine saved gas and helped me shake off my cabin fever. The old sidewalks along River Drive led me past a vine-covered stone pillar with a corroded metal plaque attached. One day, out of curiosity, I brushed aside the dirt and the vines to discover that it was a historic monument commemorating the location of the very first train bridge ever to span the river. The pillar was constructed with stones from the bridge's original abutments. Suddenly it made sense that the street that descended the bluff near my house was named "Bridge Street," though no longer was there a bridge there.

But the old marker would serve as a reminder to me on every walk thereafter that there had to be a "first of a kind" of everything, and the very first rail bridge anywhere on that 2300-mile riverway had been anchored a stone's throw from my apartment. One day a young lawyer named Lincoln would arrive in Davenport to represent the train company that had erected the structure. He won his case, so it would be established in law that a water navigation route could legally be traversed by a railroad bridge. That set a precedent in the 1800's and I would come to believe that it was not happenstance that I lived at that spot. The monument spoke to me every time I passed, reminding me that if God wants me to accomplish something – even something extremely unlikely to succeed – even something that could set a precedent – there is nothing arrayed against him powerful enough that he could not overcome it.

One morning I climbed the steps at the Post Office and turned the key in the box to discover a letter from Broward County in Florida. Contained within was a bid request from the Fort Lauderdale International Airport for a radio system of just the kind I hoped to sell. How had they found my kitchen table business with no internet search engines? It wasn't even listed in the phonebook.

But I still had my Honda. My ladder and my tools were still in the garage. I knew my equipment costs by heart and it was simple to add travel expenses to the recipe. I filled out the bid form and slid it back into the mail slot. This was totally new territory. I had just submitted a bid response in the new company's name with legal, financial and practical consequences if I failed to perform.

Then the wait began again.

7 Years Completed. One of the leaf-rakers called and invited me to a house party which I attended with great expectations of meeting someone new. Though I saw some people who I wished I could have become acquainted with, alas the circle I found myself in was not so interesting.

But a few days later I received a better call. It was from the Purchasing Department in Broward County. They were ready to accept my offer despite a protest from an established vendor who had submitted a much higher entry. The competitor was claiming that based on my low quote, my company could not possibly understand the project requirements. But my written response had promised that I would provide the specified products and services at the offered price. It was right there in black and white. And my bid was significantly lower. So, the real reason for their call? I came to understand that the County was calling just to see if I would provide them with confidence that I understood what was required ... and that I would commit to doing it. It was already written in my bid response but they wanted to hear me say it.

This was by far my biggest "as if" moment.

In broadcasting, it's a given that a convincing delivery sells products and pays bills. We routinely employed "words without knowledge" to make that happen. But I was beginning to understand through the Lord's Spirit, a whole new level of responsibility was required. From now on, my words must be "with knowledge" – the knowledge that I am being observed by a father who will absolutely hold me to account if the truth gets bent.

I told the County in no uncertain terms that my company would perform. A few days later I received a letter informing me that they had dismissed the protest ... and the airport contract was mine for a signature.

Gulp.

The Possible Impossible

Those seven years on the banks of the big river were like the bridge monument; they were most significant when viewed in hindsight. It was after that time was completed that I began to sense a foundation firming up beneath me.

Perhaps I should have anticipated it when I arrived in Iowa the fourth month of 1977. Should I have foreseen that "4-7-7-7" would signify "the fourth month of '77 plus 7 more years?" Or did I have to live the time first so I could look back and divine that? Of course, only God knows. But it was after that time passed that the impossible occurred.

I drove to Fort Lauderdale with my ladder-laden Honda to perform the first official installation in the company's new name. I was aware that my travel rig looked totally back-woods. The ladder was way longer than the car. At one gas station down south, a man spotted my Iowa plates and walked over to ask me if I had driven all the way down there to elope.

Interestingly, the trip coincided with an eclipse of the sun and I timed the itinerary so I could pull off south of Atlanta just as the moon's shadow brushed across the landscape. As darkness approached, you could feel the wind begin to cool; birds began to sing; limbs on the trees rustled noisily. Moments later it was all over and I was driving south again.

The light of day returned.

The radio transmitter from the Pennsylvania company was of higher quality than the one from Minnesota. It worked flawlessly and the airport got the service they were hoping for. In fact, it all came together so well that I caught myself dreaming that someday I might have a work truck, better tools and maybe even a real office with an employee to help.

On my return, I made a detour to visit the girl with the camera. She had married a great guy and moved to the Sunshine State three years before. She and her husband and I had maintained our friendship and committed to continue the camping tradition each summer at Big Sable Point. But in a way, they had been with me all along - even at that mountain-top moment on the sand dune. And though I don't know it for a fact, I believe that I am indebted to them for praying for me during those dark days. I still have the object she put into my hand – a small metal cross she hastily removed from her necklace, on which is inscribed "Jesus Christ is Lord."

Continuing north through Tennessee, I caught myself thinking that if the business grew, perhaps I might even meet a woman who wanted a man who could earn an honest income. I was beginning to understand that with God's help, perhaps I could turn from my old ways and become a man like the one who had lived next door.

When I arrived back in Iowa, there was an invitation in my mailbox to another group event and – I know this is absolutely going to sound scripted – a girl that I had glimpsed at the house party showed up. I chanced to ask her if she had ever been sailing. She agreed to go to the lake with me the very next day.

I sensed that the winds would be just right.

I soon discovered that this girl wasn't at all nervous about boats. In fact, she seemed to be really at home around water, even wearing her bathing suit under her shirt and shorts so she could be ready for a swim. She explained that she had been on a swim team and had even piloted boats like mine when she was growing up. I told her that I was not originally from Iowa, but I had hailed from Ohio in my early days.

"Last week a pastor and his wife from our church moved to Ohio," she offered. Then she casually added, "They moved to a little town there – named Xenia."

Now I'm pretty sure you are convinced that I'm making all this up, but I am not. It really happened. And all the while I was learning that just like with the new radio transmitter, God's Spirit knew what I needed before I did.

Then came the epic moment: out of the blue the girl announced, "It's really hot out here." Without another word she rolled backwards off the boat deck and splashed into the water, heels-over-head. More than surprised, I turned the boat around to pick her up. She was laughing as she climbed back in. Hardly an accident, she clearly loved water. Since the wind was slacking, I dropped the sail so we could swim and cool off. Her hair was everywhere and so – like girls do – she dunked her head and tilted it back as she surfaced, allowing her hair to train away from her face.

When her eyes opened, I saw them. Each one had something like a tiny gem reflecting the iridescent blue of the summer sky. There were tiny white lines. As they danced, I tried not to stare.

That is when the grains of sand began falling away.

The more we talked, the more I began to understand. The dream was all about the pressure she had endured – pressure from the very people in her life she had most needed to trust. It could so easily have crushed her, but her faith had diverted the disappointment to forge, instead, glowing gems. Like a pair of sapphires in a jeweler's case, they were right there for anyone to see.

A year after that sailing trip, Megan finished her accounting degree and assumed her role as our fledgling company's financial manager. It was her input that would spur its steady growth. In another year we married and moved our tiny company to Western Michigan where it was incorporated with her as the corporate secretary. Against all the odds, the firm survived that recession of the early 80's and continued on for more than 40 years, providing reliable revenue that supported us, Megan's daughter and our three children. The benefits of that blessing now flow to our children's children, and even to our employees and their families.

And yes, we have seen many, many circuit boards in that time.

How could it all have started with a campfire, a kitchen table and a Sunday afternoon sailing trip? How could the origins of both our family and our company be drawn back to that seventh year? No, it's not coincidental. It is only possible if behind the scenes there's a director actively calling the shots on our behalf; not a deceptive Christof, but a transparent, loving father who gently invites us to see reality.

He stands at the stage door and patiently knocks.

Consider for a moment a supreme being capable of calling into existence an entire universe of exquisite detail and proportion. How difficult would it be for him to manage the details within his universe? If you think of it that way, perhaps it's a relatively minor undertaking for God to arrange the circumstances of our lives. There are so many examples in the scriptures of him allowing a good - or a bad thing - to happen to encourage, test, teach or admonish, all depending upon what is required at the moment. How he does so while *also* extending to us free will is a mystery; one I don't think man will ever understand. But even though we will never wrap our minds around the "how" and the "why," Job continues to call to us through the ages, counselling us to be thankful for it all.

Now I will guarantee you that I really didn't understand just how involved our father is in our minute-to-minute affairs until I thought about Gary one day. Remember sandy-haired Gary from my university dorm? The year after Megan and I were married, I was marveling at the changes that had so suddenly come into my life. My mind pulled up an image of Gary, smiling as he passed me on the stairs. I really never knew him, but it was so obvious even then that his kindly nature was inspired. I wondered, "What would Gary say if I told him that I had come to believe as he did? Perhaps I should try and locate him to tell him personally." I actually considered it for a moment. But exactly how would I do that? In that day, there was no internet; no searchable media. Perhaps I could have contacted the University, but I didn't even know what city or state Gary was from. I couldn't even recall his last name. So, I dismissed the notion.

Then, a few months later when I was on my way to Seattle, I had to do a plane change at the Minneapolis airport. Before cellphones were invented, airports had walls with long rows of land-line telephones. Business people got off the planes and went right for the nearest one.

As I disembarked, I spied that only one phone was available. I was about ten paces away when I noticed that the man on the phone immediately to the right of the one that would be mine was very tall and very lanky. He had a large mop of sandy hair. As I approached, I could hear the man speaking. Instead of making a phone call, I only pretended. I was just close enough to hear the man's voice. In my other ear, the phone recording repeated, "If you'd like to make a call ...." Meanwhile, my mind was repeating "That can't be this Gary from Indiana. It *does* look a lot like him. But, exactly how am I going to know for sure? He's not going to just state his first and last name out loud while I am eavesdropping!"

Then, as if on cue, the man did just that, saying, "Yes, please tell him that Gary Stone called."

"Yes, that was Gary's last name - *'Stone.'* His name was *Gary Stone!"*

At that moment he hung up. I pretended to hang up too, and as he picked up his bag to walk away, I reached out and touched him on the shoulder. He looked around and somehow, I was able to utter, "You probably don't remember me from Indiana, do you, Gary?"

I was about to be really embarrassed because up-close, the man didn't look like the Gary Stone that I remembered. But he stopped and turned and stared right at me for a moment. His expression suddenly softened. He smiled broadly, nodded and said, "Yes, Bill. I surely do remember you. How in the world have you been?"

It was his look that said it all. It was the same acknowledging look he would give me when I passed him on the stairs. But his instant recollection of my name – how did he even recall our glancing acquaintance? I reached out and shook his hand and said, "Gary, I just want you to know that I eventually came to believe just like you did. I really didn't understand any of it back then when we were in college."

A satisfied look radiated across Gary's entire countenance. He set his travel bag down right in the middle of that busy airport aisle with passengers bumping us left and right. His blue eyes welled up. He just reached out and shook my hand again. He said only one word: "Yes." Then he wiped his eyes with the back of his hand, smiled and continued on his way.

On the plane to Seattle, I wondered if Gary thought that I might have sought him out. Or did he think our meeting was accidental? Then I thought better of it. Gary certainly knew way better than me that the Almighty is capable of pulling two people who barely knew each other in the distant past to the same airport, to adjacent payphones at the very same moment in time; even allowing them to recognize each other and recall each other's names. It was one more impossible scenario that was hardly an impossibility.

But what was the point?

My guess is that this was a very personal demonstration to me – and perhaps to Gary – and now to you – of our father's sheer, unrestrained power. This unlikely meeting made two things clear to me: 1) God's Spirit is always active, listening and seeking the best for me, even when I am not aware, and 2) there is absolutely nothing that he cannot cause to happen – and correspondingly – there is absolutely nothing I can do to impede him if he wills that something should take place.

I was reminded of the time when Jesus advised his disciples, "With God, *all* things are possible." (76) And that would explain why the Old and New Testaments are filled cover-to-cover with events that had never occurred before and have not occurred since:

_ a person walking on the surface of a lake.

_ the commanding of a thunderstorm to cease.

_ the creation of a dry path across a major body of water.

_ a woman giving birth with no interaction from a man.

_ the instant healing of a person who was born blind.

_ the feeding of thousands of people when there was no source of food

_ the stopping of the progress of the sun in the sky.

_ a person speaking in one language being understood by those who only understood other languages.

_ darkness occurring at noontime.

_ a man raising another man from the dead.

_ a man raising *himself* from the dead.

Truly, if God wills something, nothing is impossible. He has proven this to us a thousand ways and a thousand times.

And since that is the case, he is therefore capable of doing the unimaginable: raising you and me back to life with him after our own deaths. Jesus tells us that he was merely the first to rise, and he invites us to follow. Remember that he affirmed, "I am *the way*, the truth and the *life*."(8) And he promises that he will give us this new life permanently if we just trust what he tells us. "Anyone who believes in me will live, even though they die." (80)

As was demonstrated to me, you don't have to wait for your last breath. The benefits begin the moment you believe him. The new spirit born within you is your proof and your guarantee; your guide and your passport. You will look over your shoulder and see that you've made a critical turn onto a much brighter road. Far from being perfect, you will find yourself less and less tethered to the destructive and more and more bound for the constructive. But if you are like me, you will more often find yourself looking forward than back.

My cousin Judy was hoping to read what you are reading right now...but it did not happen. She took her last breath today as I wrote these very words. Judy was a blessing and an encouragement and she is already able to glimpse her future. I will dedicate this book in memory of Judy, who trusted what God said.

I have prayed specifically that you can accept these truths. Perhaps you are thinking, "How can this writer – a stranger – be praying for me? How could he know that I am going to read his book?" I am quite certain that God knows the name of every person who will read this sentence throughout time, and it is through his omnipotence that I direct my prayers on your behalf. He knows that you have this book in your hands at this very moment, just as he knew where to find Gary Stone so our divergent paths would cross. Because our father exists both inside and outside of space and time, our prayers to him are not bounded by geographies and dates.

I believe that's what happened with my grandparents Willie and Louise. The next day dawned and Willie's sinus operation commenced. The anesthetic was administered and he "went to sleep." Then something unfortunate happened. Medical professionals I have talked to speculate that his heart may have been weakened by years of tobacco use. And back in the 1940's, the administering of anesthetics was not a refined science. All I know for sure is that during the surgery, Willie went into cardiac arrest. Then he was successfully resuscitated.

Horrified, Jessie and her husband raced back to Memphis. My father Billy hitched a military flight and was met by the family at the airport. Back at the hospital, the family was told that the surgery would not be completed until Willie's condition stabilized.

It never did.

When Willie crossed the big river the final time it was in a hearse. He arrived to a shocked and saddened Forrest City. A street in the town would even be named in his memory. Judy was a child in attendance at his funeral. She was left to ponder our grandfather's story for a lifetime. Born a decade later, I would only hear the echoes.

Despite some similarities, Billy and Willie could not have been more different. That was underscored when my mother surprised me one day with their personal Bibles. Billy's was an Army-issue version in mint condition. The binding snapped. It appeared that it had never been opened. His father Willie's was tattered and dog-eared; its pages yellowing; the cover was coming off. Clearly it had been read almost to the point of disintegration.

Then I noticed that pressed between the pages of Willie's well-worn copy was what appeared to be a bookmark. On a closer look, it turned out to be a poem clipped from a newspaper:

The Bridge Builder

By Will Allen Dromgoole

"An old man going a lone highway,
Came, at the evening cold and gray,
To a chasm vast and deep and wide.
Through which was flowing a sullen tide
The old man crossed in the twilight dim,
The sullen stream had no fear for him;
But he turned when safe on the other side
And built a bridge to span the tide.

'Old man,' said a fellow pilgrim near,
'You are wasting your strength with building here;
Your journey will end with the ending day,
You never again will pass this way;
You've crossed the chasm, deep and wide,
Why build this bridge at evening tide?'
The builder lifted his old gray head;
'Good friend, in the path I have come,' he said,
'There followed after me to-day
A youth whose feet must pass this way.
This chasm that has been as naught to me
To that fair-haired youth may a pitfall be;
He, too, must cross in the twilight dim;
Good friend, I am building this bridge for him!'" (77)

Perhaps after Willie and Louise read this this poem, they prayed it would be Billy who would one day turn and follow in faith. He never did. But instead, he married a woman who years later would take their son into a Catholic Church, where the boy heard enough of God's words to become curious.

Perhaps it's a coincidence that Willie's Bible with the old poem came into my possession? I really don't think so. You may have guessed already that discovering that poem prompted me to put pen to paper. I have never written a story until now, much less a book, which further proves that the words you have found here were inspired and not invented.

Truly, they were placed here for you.

The Road To Find Out is winding. It follows the river all the way down to its end. The way is wide and shady and the walking is easy. There is no want for company. But at the river's mouth no one can proceed. I last saw my dad at the awful moment he discovered he was out of options. There was no time for turning back. So, retrace your steps. There was a turn that led to a crossing; a new way that's very narrow but very well-marked.(78)

Thankfully, our father is patient. He knows that to overcome is a struggle. He knows that to change course will mean not walking with the crowds. But be thankful he has provided for you, and for me, time to turn back; time to avoid my dad's dead-end.

Epilogue

I am honored you have taken time to flip through these pages. I am sorry if it reads like an autobiography in places, but I needed to put it into story form and mine is the only story I know. Of course, the real story isn't about me. It's about the one who overcame on our behalf.

A person's core essence consists of what he or she believes. That explains why those who desire the truth are drawn to God's wisdom and those who thrive on lies, to a self-indulgent wasteland. (98) If this little book can claw even one out of the web of untruth, it's been worth writing it.

You may have noticed that I do not have any letters beside my name. The things written here are not intended to be a scholarly treatment of Biblical writings. And the point is not to suggest that if you believe what God says, "impossible" occurrences are going to happen to you. They might. But more likely, the things in this book have happened in order to get my attention. Being a "math and science guy," I have always noticed those high-odds situations, and I apparently needed that "extra" to shout at me loudly and frequently.

In so many corners of our world, God's words are not welcome. In some places they can get you imprisoned or killed. The truth will always be a threat to authoritarians and tyrants because their hold on power is tenuous. It's built on a foundation of lies. So, if this writing is all you have, keep these words within you as you proceed. As the Apostle John told us, "In him is life and the life is the light of all men. The light shines in the darkness and the darkness did not overcome it." (79) Indeed, they can take away your books and take away your life, but not even the evil one himself has the power to eclipse your faith. So, hide God's words within you. They will bring you life that overcomes even your own death. Remember, anyone who believes God will live, even though they die. (80)

If you are fortunate enough to have access to the scriptures, the Book of John is an excellent place to begin reading. And in the End Notes that follow, I have included the references for the various verses cited here, so you can look them up in their contexts.

If you are Catholic, my comments and illustrations are not meant to demean you or suggest that you don't know the truth. They are meant solely as a criticism of the Catholic theology that I was introduced to.

No, there is no such thing as an accident. You know why. An accident is an event that no one can anticipate. But since God knows all things prior to them happening, nothing can occur accidentally. You were intended to read this. Knowing that God directed this book into your hands, I encourage you to find another who can partner with you, so you will both grow faster in faith. A partner will also help you remain strong when you are tempted to trespass in word or deed; to disregard God's design for life. Do what I should have done. Find your Gary Stone. You may already know who he or she is. While there is time, get together and pray for each other in the Lord's presence. We are told that "a cord of three strands is not easily broken." (81)

A loyal partner can be found in a church if you use discretion, understanding that there are some who profess faith but truly do not have any. Jesus told us this would be so. He himself did not have to look far to see a play actor in his immediate company – a fraud who feigned faith and discipleship, but in reality, loved only money. Judas, was one of the twelve. He pretended to be a kinsman, but he was really a killer. If he had taken Jesus' teachings to heart, he would have heard him advise, "No one can serve two masters, for either you will hate one and love the other, or else you will be devoted to one and despise the other. You cannot serve both God and worldly wealth." (82) Judas made the biggest miscalculation in the history of miscalculations when he traded the only one who could save him for a meager pay-out he couldn't keep.

So, beware of one who claims to be a brother or a sister but practices materialism. It's a killer. "Be on your guard against all kinds of greed," Jesus cautioned. "Life does not consist of an abundance of possessions." (83) Materialism is founded on a lie, and like all lies, it is not merely harmful, it is a malignancy. Lies not only kill, they spread as they do. That is why Jesus compared lies to yeast; a pinch quickly travels through an entire lump of dough. Lies and slander spread from their host to infect all their relationships. (97) Untruth may even cloak itself with an allurement that beguiles with images of the future or the past. But watch carefully: if God's name is invoked, it will bolt abruptly; it obeys an authority with no appetite for the truth.

If reaching out to others is too dangerous, be patient. As you get to know your father one-on-one, do the best you can with the resources he has entrusted to you. When the time is right, invite others to look past the reality they have been presented; to trust what God says and overcome.

I finished writing this beneath a stone archway that was once the entrance into an old school building in Indiana. It would grow to become the university from which I graduated. As I was researching its history, it became apparent that along the way its leaders had undergone a radical transformation. The first class in 1825 had just ten students, but there came steady growth. Then, after the Civil War, a fire prompted them to relocate to the outskirts of town where a new campus would be carved from a parcel of woodland purchased from a local farmer.

But what would become of the original building, now surrounded by a growing city? On it were the beautiful stone portals through which its first students walked to class; attached to the building's exterior was the first official seal of the institution. Their novel solution was to remove the arches and the seal and bring them to the new property. Workers integrated them into an 8-sided gazebo, in the center of which they installed a hand-operated water pump. The new structure was christened the "Well House," and affixed to its exterior – still to this day – is that original seal.

When I was a student I would walk past the Well House, never giving a thought to its history. But had I looked carefully, I would have spotted the old seal, still in evidence on the west side of the octagon. If I had taken the time to compare, I would have noticed that it is distinctly different from the modern renderings. All of them include an image of an open book and the words "Lux Et Veritas," in Latin, "Light And Truth." But only the first seal features the words "Holy Bible" emblazoned across its pages. Modern versions still retain the open book icon, but now the pages display only nonsensical dashes to represent words. Once it was understood that all knowledge is based on God's words. Today, however, we are presented with a radically different proposition: the Good Book has been replaced by a generic one. Symbols that bear no meaning populate its pages.

Could there be a better illustration of "words without knowledge?"

Standing in that Well House one day, it occurred to me that generations of students – including myself – have been intentionally misinformed. My school had once acknowledged publicly that God's words constitute every person's Light and Truth. The seal stated unequivocally that his words are the very cornerstone of human knowledge. But apparently, what was self-evident when the institution was birthed, went absent as it matured. There is no reference to the change in the school's published history, but someone in the administration must have decided that his words are no longer paramount. The words "Holy" and "Bible" did not just fall off the school's seal. They were erased intentionally, and certainly with the approval of the university's leadership. As a result, many students like me came and left that institution continuing to believe that what God says is irrelevant.

I point this out as a caution: we have a formidable adversary, but he is not often arrayed as we would expect. The evil one takes root in the state that's isolated from the church. He holds sway where holy is homogenized and God's name is unwelcome. He dominates where the Creator's designs are disrupted and anarchy is elevated.

Evil triumphs when the words of truth from the Good Book are replaced with gibberish from a generic one. And the yeast always spreads. As I saw firsthand in my brief media career, any institution that refuses to acknowledge the author of life, ultimately will disseminate only death.

So, I hope you will exercise the greatest care as you walk your road. I started my journey like a student driver. I was hesitant; afraid of making mistakes. But I found that the best driver is the one who designed the car. Only after I trusted him to take the wheel did my life begin to go somewhere. Then the changes did come quickly. Across the five-years that began the day I believed, I was changed from an unemployed radio announcer to a married man with children, a house and a growing business. I often didn't know where I was going, day to day. I just trusted that the driver knew the way.

Thankfully we are all presented with life one moment at a time. If we were aware, in advance, of all the future troubles we will encounter, who would want to proceed? But I have found that just being certain of the destination makes all the difference.

It's incredible to me that this book now contains more than 50,000 words. Although it may seem as if it is one piece of writing, in fact, it was stitched together from a thousand pieces, doled out a few at a time: one on a plane; another at a hotel room; a few more before bed. Once I booked a cottage in Indiana with my hiking friend. When he took ill and couldn't make the trip, the two days I spent there without him provided the opportunity to make corrections. When I went to leave, I spotted the road sign and bridge that are pictured on the cover. They were right outside the cottage's window. Ironically, my friend's cancellation had provided the time and place to complete the changes and capture the image. And when Megan and I did the research for this book's End Notes, we were quarantined - stranded in the mountains of Wyoming as we waited to get well. It took those quiet hours of uninterrupted time to get all the citations researched accurately.

Though the process that created what you are reading was as unpredictable as the wind, the result was never in doubt.

The boldest demonstration of this I witnessed while working one day in California. The site was a lonesome highway interchange along Interstate 5 – the heavily-travelled conduit that links the Bay Area and Los Angeles. That particular exit did not have any services, so I was depending totally on my cooler for food and drink that day. As the sun began to swing low, a hitchhiker appeared out on the entrance ramp. He was trying to catch the eye of someone heading south.

Putting away my tools, I noticed he was still thumbing, even as dusk was fading. The streaks of the headlights and taillights were all that allowed me to make out his silhouette. I thought, perhaps, I could give him a ride to a hotel before it got much later, so I walked down the ramp to see.

The hitchhiker was hardly the rough character I was expecting. In fact, I would rank him among the most kindly and gracious strangers I have ever encountered. He told me that he was on his way to meet some friends in Los Angeles, and together they would be attending a conference of faithful people there.

I expressed that he might not find a ride at that hour. It was totally dark. No one could see him. But he did not appear to be in the least worried. Amidst the roar of the traffic and the swirling desert wind, the man turned to me and said calmly, "It's okay, my friend. I know exactly where I am going."

The Bridge Builder

An old man, going on a lone highway,
Came at the evening, cold and gray,
To a chasm vast and deep and wide.
Where no bridge stretched from side to side
The old man crossed in the twilight dim;
The swollen stream had no fear for him.
But he turned when safe on the other side
And built a bridge to span the tide.

"Old man," said a fellow pilgrim, near,
"You are wasting your strength with building here;
Your journey will end with the ending day;
You never again will pass this way;
You've crossed the chasm deep and wide.
Why build you this bridge at eventide?"

The builder lifted his old gray head:
"Good friend, in the path I've come," he said,
"There followeth after me today
A youth whose feet must pass this way.
This chasm that has been as naught to me,
To that fair-haired youth may a pitfall be;
He, too, must cross in the twilight dim,
Good friend, I'm building this bridge for him."

Will Allen Dromgoole, poetess of the Nashville Banner for many years before her death.

Thank You

Thank you to Megan, Malinda, Linda, Athena, Cyndy, Vicki and Jack, Robert, Rich, Greg and Sharyl, Martha, Ken, Allan, Ross and Wayne for your encouragement, your editing eyes and your amazing advice.

Special thanks to Vicki, Megan, Steve, Mary Lou, Willie and Louise, Billy, Rose, Jessie, George, Judy, Tom D, Frank E, Greg, Tom, Gary, Jeff L, Morana, John, Spike, Jeff C, Rich, Sue, Kathy and the twilight hitchhiker for the use of your stories.

And an extra special thanks to Malinda for her "striking" oak tree artwork.

You can also download *The Road To Find Out* as a free ebook or obtain a hard copy at a variety of vendors listed at:

books2read.com/u/bPLeKY.

To support this nonprofit effort, inquire at:

OVERCOMEInitiative@gmail.com

All revenue realized through donations or the sale of this writing will be used to produce books for distribution at no charge.

End Notes

All references are taken from the New King James version of the *Bible.*

1. Job 38:2-6, 31.
2. Genesis 1:1-3.
3. Genesis 3:20.
4. Genesis 2:1.
5. John 11:10.
6. Proverbs 1:7.
7. Matthew 25:34-40.

   "Then the King will say to those on His right hand, 'Come, you blessed of My Father, inherit the kingdom prepared for you from the foundation of the world: for I was hungry and you gave Me food; I was thirsty and you gave Me drink; I was a stranger and you took Me in; I was naked and you clothed Me; I was sick and you visited Me; I was in prison and you came to Me.' Then the righteous will answer Him, saying, 'Lord, when did we see You hungry and feed You, or thirsty and give You drink? When did we see You a stranger and take You in, or naked and clothe You? Or when did we see You sick, or in prison, and come to You?' And the King will answer and say to them, 'Assuredly, I say to you, inasmuch as you did it to one of the least of these My brethren, you did it to Me.'"
8. John 14:6.
9. John 7:45.
10. John 7:46.
11. Matthew 6:25-33.
12. Acts 17:22-29.

    "Then Paul stood in the midst of the Areopagus and said, 'Men of Athens, I perceive that in all things you are very religious; for as I was passing through and considering the objects of your worship, I even found an altar with this inscription: TO THE UNKNOWN GOD. Therefore, the One whom you worship without knowing, Him I proclaim to you: God, who made the world and everything in it, since He is Lord of heaven and earth, does not dwell in temples made with hands. Nor is He worshiped with men's hands, as though He needed anything, since He gives to all life, breath, and all things. And He has made from one blood every nation of men to dwell on all the face of the earth, and has determined their pre-appointed times and the boundaries of their dwellings, so that they should seek the Lord, in the hope that they might grope for Him and find Him, though He is not far from each one of us; for in Him we live and move and have our being, as also some of your own poets have said, 'For we are also His offspring.' Therefore, since we are the offspring of God, we ought not to think that the Divine Nature is like gold or silver or stone, something shaped by art and man's devising.'"

13. Genesis 1:26.

"Then God said, 'Let Us make man in Our image, according to Our likeness; let them have dominion over the fish of the sea, over the birds of the air, and over the cattle, over all the earth and over every creeping thing that creeps on the earth.'"

14. I John 4:7-8.

"Beloved, let us love one another, for love is of God; and everyone who loves is born of God and knows God. He who does not love does not know God, for God is love."

15. John 1:1-3.

16. John 5:19-23.

17. John 15:26.

18. Isaiah 55:8.

19. Psalm 18:30.

"*As* for God, His way *is* perfect; The word of the LORD is proven; He is a shield to all who trust in Him."

20. Romans 3:20

"Therefore by the deeds of the law no flesh will be justified in His sight, for by the law is the knowledge of sin."

21. Matthew 22:37-40.

"Jesus said to him, 'You shall love the Lord your God with all your heart, with all your soul, and with all your mind.' This is the first and great commandment. And the second is like it: You shall love your neighbor as yourself. On these two commandments hang all the Law and the Prophets.'"

22. Luke 18:9-14.

23. James 2:19.

"You believe that there is one God. You do well. Even the demons believe—and tremble!"

24. Romans 1:20-21

"For since the creation of the world His invisible attributes are clearly seen, being understood by the things that are made, even His eternal power and Godhead, so that they are without excuse, because, although they knew God, they did not glorify Him as God, nor were thankful, but became futile in their thoughts, and their foolish hearts were darkened."

25. John 8:31-32.

26. Luke 2:10-11.

27. Luke 1:39-45

"Now Mary arose in those days and went into the hill country with haste, to a city of Judah, and entered the house of Zacharias and greeted Elizabeth. And it happened, when Elizabeth heard the greeting of Mary, that the babe leaped in her womb; and Elizabeth was filled with the Holy Spirit. Then she spoke out with a loud voice and said, 'Blessed are you among women, and blessed is the fruit of your womb! But why is this granted to me, that the mother of my Lord should come to me? For indeed, as soon as the voice of your greeting sounded in my ears, the babe leaped in my womb for joy. Blessed is she who believed, for there will be a fulfillment of those things which were told her from the Lord.'"

28. Luke 1:11-20; 57-64.

29. Luke 1:26-38.

"Now in the sixth month the angel Gabriel was sent by God to a city of Galilee named Nazareth, to a virgin betrothed to a man whose name was Joseph, of the house of David. The virgin's name was Mary. And having come in, the angel said to her, 'Rejoice, highly favored one, the Lord is with you; blessed are you among women!' "But when she saw him, she was troubled at his saying, and considered what manner of greeting this was. Then the angel said to her, 'Do not be afraid, Mary, for you have found favor with God. And behold, you will conceive in your womb and bring forth a Son, and shall call His name Jesus. He will be great, and will be called the Son of the Highest; and the Lord God will give Him the throne of His father David. And He will reign over the house of Jacob forever, and of His kingdom there will be no end.' "Then Mary said to the angel, 'How can this be, since I do not know a man?' "And the angel answered and said to her, 'The Holy Spirit will come upon you, and the power of the Highest will overshadow you; therefore, also, that Holy One who is to be born will be called the Son of God. Now indeed, Elizabeth your relative has also conceived a son in her old age; and this is now the sixth month for her who was called barren. For with God nothing will be impossible.'

"Then Mary said, 'Behold the maidservant of the Lord! Let it be to me according to your word' And the angel departed from her."

30. Genesis 12:1-8.

"Now the Lord had said to Abram: 'Get out of your country, from your family and from your father's house, to a land that I will show you. I will make you a great nation; I will bless you and make your name great; and you shall be a blessing. I will bless those who bless you, And I will curse him who curses you; and in you all the families of the earth shall be blessed.' "So, Abram departed as the Lord had spoken to him, and Lot went with him. And Abram was seventy-five years old when he departed from Haran. Then Abram took Sarai his wife and Lot his brother's son, and all their possessions that they had gathered, and the people whom they had acquired in Haran, and they departed to go to the land of Canaan. So, they came to the land of Canaan. Abram passed through the land to the place of Shechem, as far as the terebinth tree of Moreh. And the Canaanites were then in the land. "Then the Lord appeared to Abram and said, 'To your descendants I will give this land.' And there he built an altar to the Lord, who had appeared to him. And he moved from there to the mountain east of Bethel, and he pitched his tent with Bethel on the west and Ai on the east; there he built an altar to the Lord and called on the name of the Lord. So Abram journeyed, going on still toward the South. "Now there was a famine in the land, and Abram went down to Egypt to dwell there, for the famine was severe in the land. And it came to pass, when he was close to entering Egypt, that he said to Sarai his wife, 'Indeed I know that you are a woman of beautiful countenance. Therefore, it will happen, when the Egyptians see you, that they will say, 'This is his wife'; and they will kill me, but they will let you live. Please say you are my sister, that it may be well with me for your sake, and that I may live because of you.' "So it was, when Abram came into Egypt, that the Egyptians saw the woman, that she was very beautiful. The princes of Pharaoh also saw her and commended her to Pharaoh. And the woman was taken to Pharaoh's house. He treated Abram well for her sake. He had sheep, oxen, male donkeys, male and female servants, female donkeys, and camels. "But the Lord plagued Pharaoh and his house with great plagues because of Sarai, Abram's wife. And Pharaoh called Abram and said, 'What is this you have done to me? Why did you not tell me that she was your wife?'"

31. Genesis 13:14-15.

    "And the Lord said to Abram, after Lot had separated from him: 'Lift your eyes now and look from the place where you are—northward, southward, eastward, and westward; for all the land which you see I give to you and your descendants forever.'"

32. Genesis 15:4.
33. Genesis 15:5.
34. Genesis 15:6.
35. Numbers 21:9.
36. John 3:14-15.
37. John 3:9-13.

    "Nicodemus answered and said to Him, 'How can these things be?' Jesus answered and said to him, 'Are you the teacher of Israel, and do not know these things? Most assuredly, I say to you, We speak what We know and testify what We have seen, and you do not receive Our witness. If I have told you earthly things and you do not believe, how will you believe if I tell you heavenly things? No one has ascended to heaven but He who came down from heaven, that is, the Son of Man who is in heaven.'"

38. Genesis 2:16-17.

"And the Lord God commanded the man, saying, 'Of every tree of the garden you may freely eat; but of the tree of the knowledge of good and evil you shall not eat, for in the day that you eat of it you shall surely die.'"

39. Genesis 3:1.

40. John 8:44.

"You are of your father the devil, and the desires of your father you want to do. He was a murderer from the beginning, and does not stand in the truth, because there is no truth in him. When he speaks a lie, he speaks from his own resources, for he is a liar and the father of it."

41. Genesis 3:5.

42. Luke 15:11-24.

"Then He said: 'A certain man had two sons. And the younger of them said to his father, 'Father, give me the portion of goods that falls to me.' So he divided to them his livelihood. And not many days after, the younger son gathered all together, journeyed to a far country, and there wasted his possessions with prodigal living. But when he had spent all, there arose a severe famine in that land, and he began to be in want. Then he went and joined himself to a citizen of that country, and he sent him into his fields to feed swine. And he would gladly have filled his stomach with the pods that the swine ate, and no one gave him anything. "But when he came to himself, he said, 'How many of my father's hired servants have bread enough and to spare, and I perish with hunger! I will arise and go to my father, and will say to him, 'Father, I have sinned against heaven and before you, and I am no longer worthy to be called your son. Make me like one of your hired servants.' "And he arose and came to his father. But when he was still a great way off, his father saw him and had compassion, and ran and fell on his neck and kissed him. And the son said to him, 'Father, I have sinned against heaven and in your sight, and am no longer worthy to be called your son.' "But the father said to his servants, 'Bring out the best robe and put it on him, and put a ring on his hand and sandals on his feet. And bring the fatted calf here and kill it, and let us eat and be merry; for this my son was dead and is alive again; he was lost and is found.' And they began to be merry."

43. Matthew 10:38

"And he who does not take his cross and follow after Me is not worthy of Me."

44. Genesis 3:15.

45. Isaiah 7:14.

46. Psalms 89:3-4.

47. Micah 5:2.

48. Isaiah 53:3.

49. Matthew 12:39-40; Isaiah 53:10; Psalm 16: 8-11; Jonah 2:1-3:2; Hosea 6:1-2.

50. Hebrews 4:12.

51. Psalm 119:105.

    "Your word is a lamp to my feet and a light to my path."

52. John 16:33.

53. John 14:6.

54. 1 John 5:4-5.

55. 1 John 2:14.

56. Isaiah 53:6; 1 Peter 2:24.

57. John 10:18.

58. John 15:13.

59. Mark 15:39.

60. Matthew 16:24-26.

61. John 20:26-27.

    "And after eight days His disciples were again inside, and Thomas with them. Jesus came, the doors being shut, and stood in the midst, and said, 'Peace to you!' Then He said to Thomas, 'Reach your finger here, and look at My hands; and reach your hand *here*, and put it into My side. Do not be unbelieving, but believing.'"

62. John 15:26-27.

    "But when the Helper comes, whom I shall send to you from the Father, the Spirit of truth who proceeds from the Father, He will testify of Me. And you also will bear witness, because you have been with Me from the beginning."

63. Acts 2:22-40.

    "Men of Israel, hear these words: Jesus of Nazareth, a Man attested by God to you by miracles, wonders, and signs which God did through Him in your midst, as you yourselves also know — Him, being delivered by the determined purpose and foreknowledge of God, you have taken by lawless hands, have crucified, and put to death; whom God raised up, having loosed the pains of death, because it was not possible that He should be held by it. For David says concerning Him: "'I foresaw the Lord always before my face, for He is at my right hand, that I may not be shaken. Therefore my heart rejoiced, and my tongue was glad; moreover my flesh also will rest in hope. For You will not leave my soul in Hades, nor will You allow Your Holy One to see corruption. You have made known to me the ways of life; you will make me full of joy in Your presence.' "Men and brethren, let me speak freely to you of the patriarch David, that he is both dead and buried, and his tomb is with us to this day.' Therefore, being a prophet, and knowing that God had sworn with an oath to him that of the fruit of his body, according to the flesh, He would raise up the Christ to sit on his throne, he, foreseeing this, spoke concerning the resurrection of the Christ, that His soul was not left in Hades, nor did His flesh see corruption. This Jesus God has raised up, of which we are all witnesses. Therefore being exalted to the right hand of God, and having received from the Father the promise of the Holy Spirit, He poured out this which you now see and hear. For David did not ascend into the heavens, but he says himself:

"'The Lord said to my Lord, "Sit at My right hand, till I make Your enemies Your footstool."

"Therefore, let all the house of Israel know assuredly that God has made this Jesus, whom you crucified, both Lord and Christ.' "Now when they heard this, they were cut to the heart, and said to Peter and the rest of the apostles, 'Men and brethren, what shall we do?' "Then Peter said to them, 'Repent, and let every one of you be baptized in the name of Jesus Christ for the remission of sins; and you shall receive the gift of the Holy Spirit. For the promise is to you and to your children, and to all who are afar off, as many as the Lord our God will call.' And with many other words he testified and exhorted them, saying, 'Be saved from this perverse generation.'"

64. 2 Corinthians 5:17.

"Therefore, if anyone is in Christ, he is a new creation; old things have passed away; behold, all things have become new."

65. John 15:12-15.

"This is My commandment, that you love one another as I have loved you. Greater love has no one than this, than to lay down one's life for his friends. You are My friends if you do whatever I command you. No longer do I call you servants, for a servant does not know what his master is doing; but I have called you friends, for all things that I heard from My Father I have made known to you."

66. Revelation 3:20.

67. Hebrews 4:16.

"Let us therefore come boldly to the throne of grace, that we may obtain mercy and find grace to help in time of need."

68. Romans 8:26-27.

"Likewise the Spirit also helps in our weaknesses. For we do not know what we should pray for as we ought, but the Spirit Himself makes intercession for us with groanings which cannot be uttered."

69. John 3:1-3.

70. Matthew 7:22-23.

71. John 3:4-8.

"Nicodemus said to Him, 'How can a man be born when he is old? Can he enter a second time into his mother's womb and be born?' Jesus answered, 'Most assuredly, I say to you, unless one is born of water and the Spirit, he cannot enter the kingdom of God. That which is born of the flesh is flesh, and that which is born of the Spirit is spirit. Do not marvel that I said to you, "You must be born again." The wind blows where it wishes, and you hear the sound of it, but cannot tell where it comes from and where it goes. So is everyone who is born of the Spirit.'"

72. John 10:27-29.

73. 1 Corinthians 2:9.

74. John 14:3.

75. John 15:26.

"But when the Helper comes, whom I shall send to you from the Father, the Spirit of truth who proceeds from the Father, He will testify of Me."

76. Matthew 19:26.

77. Dromgoole, Will Allen. "The Bridge Builder." *Father: An Anthology of Verse*. EP Dutton & Company, 1931.

78. Matthew 7:13-14.

"Wide is the gate and broad is the road that leads to destruction, and many enter through it; but small is the gate and narrow the road that leads to life, and only a few find it."

79. John 1:4-5.

80. John 11:25.

81. Ecclesiastes 4:12.

82. Matthew 6:24.

83. Luke 12:15.

84. Matthew 7:29.

85. Psalm 34:20; John 19:33-36.

86. John 19:30.

87. Niccol, Andrew. *The Truman Show*. American Science Fiction Satirical Psychological Comedy-Drama Film. Paramount Pictures, 1998.

88. Smith, Huston. *The Religions of Man: A Clear and Objective Description of the Great Religions and Their Appeal to the Spiritual Aspirations of the Different Peoples of the World*, 1964.

89. McLean, Don, comp. *American Pie*. American Pie Album. United Artists, 1971.

90. *Sky and Telescope Magazine*. Sky Publishing Corporation, 1968.

91. Kubrick, Stanley, dir. *2001: A Space Odyssey*. Writ. Arthur C. Clarke. Metro Goldwyn-Mayer (MGM), 1968. Film. 2 Dec 2013.

92. Darwin, Charles. *On the Origin of Species: By Means of Natural Selection, Or, the Preservation of Favored Races in the Struggle for Life*, 1883.

93. Sagan, Carl, Ann Druyan and Steven Soter. "Cosmos." 13-Part Documentary Series. PBS, 1980.

94. McLuhan, M., Fiore, Q. and Agel, J. *The Medium is the Massage: An Inventory of Effects*, 1967 (p.199).

95. Warhol, Andy. 1968.

96. 1 Peter 2:9.

    "But you are a chosen race, a royal priesthood, a holy nation, a people for his own possession, that you may proclaim the excellencies of him who called your out of darkness into his marvelous light."

97. Luke 12:1.

    "Beware the leaven (yeast) of the Pharisees, which is hypocrisy.

98. Proverbs 23:7.

"As he thinks in his heart, so is he."

1. The launch of the James Web Space Telescope (JWST) in 2022 caused many Big Bang advocates to take a big step back. When the data began streaming in, results were not at all as anticipated. The universe's very earliest galaxies were unexpectedly plentiful and were observed to be much larger and more fully formed than had been predicted, especially given the very short time they would have had to "develop." It became clear that either the formulas scientists had been using to estimate the universe's expansion were way off or the "Big Bang" had not occurred.

Printed in the USA
CPSIA information can be obtained
at www.ICGtesting.com
CBHW021116161024
15764CB00005B/10

9 798223 135531